2019台达杯国际太阳能建筑设计竞赛获奖作品集

Awarded Works from International Solar Building Design Competition 2019

阳光·文化之旅
SUNSHINE & A TOUR OF CULTURE

中国可再生能源学会太阳能建筑专业委员会　编
Edited by Special Committee of Solar Buildings, CRES

执行主编：仲继寿　张　磊
Chief Editor: Zhong Jishou, Zhang Lei

编辑：鞠晓磊　夏晶晶　张星儿
Editor: Ju Xiaolei, Xia Jingjing, Zhang Xinger

中国建筑工业出版社
CHINA ARCHITECTURE & BUILDING PRESS

图书在版编目（CIP）数据

2019台达杯国际太阳能建筑设计竞赛获奖作品集　阳光·文化之旅/中国可再生能源学会太阳能建筑专业委员会编；仲继寿，张磊执行主编. —北京：中国建筑工业出版社，2019.9
　ISBN 978-7-112-24015-9

　Ⅰ.①2…　Ⅱ.①中…②仲…③张…　Ⅲ.①太阳能住宅-建筑设计-作品集-中国-现代　Ⅳ.①TU241.91

中国版本图书馆CIP数据核字（2019）第155595号

国际太阳能建筑设计竞赛是配合世界太阳能大会两年一届定期举行的常规赛事，本届竞赛以"阳光·文化之旅"为主题，通过河北兴隆暗夜公园星空驿站和浙江凤溪玫瑰教育研学基地两个赛题，面向全球征集作品，希望通过乡村良好的生态人文环境和深厚的文化内涵，以文旅融合为契机，提升乡村发展的附加值，彰显乡村的绿色发展活力。本书则为2019台达杯国际太阳能建筑设计竞赛的优秀获奖作品及竞赛过程回顾。

本书可供高等学校建筑设计相关专业本科生、研究生及建筑师参考阅读。

责任编辑：吴　绫　唐　旭　李东禧　张　华
责任校对：赵听雨

2019台达杯国际太阳能建筑设计竞赛获奖作品集
阳光·文化之旅

中国可再生能源学会太阳能建筑专业委员会　编
执行主编：仲继寿　张　磊
编辑：鞠晓磊　夏晶晶　张星儿

＊

中国建筑工业出版社出版、发行（北京海淀三里河路9号）
各地新华书店、建筑书店经销
北京雅盈中佳图文设计公司制版
北京富诚彩色印刷有限公司印刷

＊

开本：787×1092毫米　1/12　印张：22$\frac{2}{3}$　字数：678千字
2019年9月第一版　2019年9月第一次印刷
定价：188.00元（含光盘）
ISBN 978-7-112-24015-9
　　　（34513）

版权所有　翻印必究
如有印装质量问题，可寄本社退换
（邮政编码 100037）

广袤的乡村，积淀着千百年的农耕文化，浓缩了中华文明的历史变迁。经济的快速发展推动着城市化的进程，而乡村面临着文化设施匮乏、精英人才流失、文化资源未被激活和转化等问题。党的十九大提出了实施乡村振兴战略的重大决策部署，如何因地制宜地通过一系列针对性措施激活乡村资源，丰富乡村地区文化设施，让乡村文化生活迸发新的生命力也成为本次竞赛的关注点。为此，竞赛设置了研学基地和星空驿站两个赛题，希望通过乡村良好的生态人文环境和深厚的文化内涵，以文旅融合为契机，提升乡村发展的附加值，彰显乡村的绿色发展活力。

感谢台达集团资助举办2019台达杯国际太阳能建筑设计竞赛！

谨以本书献给致力于乡村建设与发展的同仁们！

The vast area in the countryside has carried thousands of years of farming culture and witnessed the historical vicissitude of Chinese civilization. The rapid economic development has driven the urbanization, but there exist such problems as a lack of cultural facilities, talent drain and inability to activate and transform cultural resources in the countryside. The ideas about major decision-making deployment for implementing the rural revitalization strategy were proposed in the 19th CPC National Congress. Specifically, this revolves around how to activate rural resources through targeted measures based on local conditions, enrich cultural facilities in rural areas, and make rural cultural life more dynamic, which has also become the focus of this competition. To this end, two research topics have been set in the competition, that is, the research base and the starry sky station. It is hoped that through the sound ecological and humanistic environment and profound cultural connotation in the countryside and with the opportunity for culture-tourism integration, the additional value of rural development will be enhanced, so as to highlight the green development vitality of the countryside.

Thanks to Delta Group for sponsoring the International Solar Building Design Competition 2019.

This book is written for those colleagues committed to rural construction and development.

目录
CONTENTS

阳光·文化之旅　SUNSHINE & A TOUR OF CULTURE

2019台达杯国际太阳能建筑设计竞赛过程回顾　General Background of International Solar Building Design Competition 2019

2019台达杯国际太阳能建筑设计竞赛评审专家介绍
Introduction of Jury Members of International Solar Building Design Competition 2019

获奖作品　Prize Awarded Works　001

综合奖·一等奖　General Prize Awarded · First Prize

风吟·花舞（凤溪）	Wind Singing · Flowers Dancing（Fengxi）	002
摘·星辰（兴隆）	Pick Stars（Xinglong）	008

综合奖·二等奖　General Prize Awarded · Second Prize

溪·聚（凤溪）	Gathered by River（Fengxi）	014
倚山居（兴隆）	Lean on the Mountain（Xinglong）	020
通往星空——河北兴隆暗夜公园星空驿站设计（兴隆）	The Way to The Starry Sky—Design of Starry Sky in Xinglong Dark Night Park in Hebei（Xinglong）	024
云汉·七居（兴隆）	The Galaxy & Dwellings（Xinglong）	030

综合奖·三等奖　General Prize Awarded · Third Prize

折院（凤溪）	Zigzag Yard（Fengxi）	034

阳光·风动（凤溪）	Sunshine and Wind Power (Fengxi)	040
予·还（凤溪）	Cycle (Fengxi)	046
伫立星河（兴隆）	Standing on the Star River (Xinglong)	052
墙褓（兴隆）	Raising in the Wall (Xinglong)	058
卧看星河尽意明（兴隆）	Lying in the Galaxy (Xinglong)	064

综合奖·优秀奖　General Prize Awarded · Honorable Mention Prize

一方（凤溪）	Nostalgia (Fengxi)	070
风檐寸晷（凤溪）	Wind Eaves Inch Sundial (Fengxi)	076
等风来（凤溪）	Waiting for the Wind (Fengxi)	082
冷巷·风院（凤溪）	Cold Lane · Wind Yard (Fengxi)	088
水院风堂（凤溪）	Water Courtyard Wind Hall (Fengxi)	094
檐下风致（凤溪）	Leisure Relaxation and Education Under the Roof (Fengxi)	100
研杭光社（凤溪）	Hangzhou Light & Sight Club (Fengxi)	106
游廊·穿院（凤溪）	Veranda · Courtyard (Fengxi)	110
光伏森林（凤溪）	Forest of Solar Photovoltaics (Fengxi)	116
云山·风居——浙江凤溪玫瑰教育研学基地设计（凤溪）	Wind House—Design of Rose Education and Research Base in Fengxi, Zhejiang Province (Fengxi)	122
游戏·园（凤溪）	Game Garden (Fengxi)	128

绿色庭院研学中心（凤溪） Green Courtyard Research Center（Fengxi）	132
呼·吸（凤溪） Breath（Fengxi）	136
方·圆 光·腔（凤溪） Round · Square Green House Cavity（Fengxi）	142
凤源·熙聚（凤溪） Sunshine Base in Fengxi（Fengxi）	146
星巢（兴隆） The Star Nest（Xinglong）	152
光·星院（兴隆） Sunlight · Watch the Stars of The Garden（Xinglong）	158
星·桥·栈（兴隆） Star · Bridge · Stack（Xinglong）	162
星光暗夜·山居秸院（兴隆） Starlight Dark Night · Hillhouse Straw Courtyard（Xinglong）	166
驿隐星升（兴隆） Relay Disappear ＆ Stars Appear（Xinglong）	172
光谷星桥（兴隆） Sun Valley · Star Bridge（Xinglong）	178
山居（兴隆） Mountain Perch（Xinglong）	184
白驹临庭 繁星入梦（兴隆） Sunshine Pouring in Courts Stars Twinkin in Dream（Xinglong）	188
归源山居（兴隆） Guiyuan Mountain Residence（Xinglong）	194
星之所向（兴隆） Direction of Stars（Xinglong）	198
落星谷（兴隆） The Valley of Falling Star（Xinglong）	202
兴隆暗夜公园星空驿站 星语心"院"（兴隆） The Starry Sky Station of XingLong Dark Night Park Dream Courtyard，Talking with Star，Wishing with Heart（Xinglong）	208
星光·叠落（兴隆） Star-Light-Sprinkle（Xinglong）	214
一期一会（兴隆） The Very Moment（Xinglong）	220
壤院（兴隆） When Earth Hugs Courtyard（Xinglong）	226

有效作品参赛团队名单
Name List of all Participants Submitting Effective Works 232

2019台达杯国际太阳能建筑设计竞赛办法
Competition Brief for International Solar Building Design
Competition 2019 242

2019台达杯国际太阳能建筑设计竞赛过程回顾
General Background of International Solar Building Design Competition 2019

2019 台达杯国际太阳能建筑设计竞赛由国际太阳能学会、中国可再生能源学会、全国高等学校建筑学学科专业指导委员会主办,国家住宅与居住环境工程技术研究中心、中国可再生能源学会太阳能建筑专业委员会承办,中国建筑设计研究院有限公司协办,台达集团冠名。在社会各界的大力支持下,竞赛组委会先后组织了竞赛启动、媒体宣传、校园巡讲、作品注册与提交、作品评审等一系列活动。这些活动得到了海内外业界人士的积极响应和参与。

一、赛题设置

开展乡村文化旅游,有效地将乡村生态资源转化成经济来源,可以促进产业兴旺、打造生态宜居环境、增强乡风文明、形成治理有效机制、最终实现生活富裕,让文化的发展创造乡村振兴的新契机。本届竞赛以"阳光·文化之旅"为主题,希望在乡村振兴背景下,进一步激活各种乡村资源,通过研学基地、驿站等多种载体,提升乡村发展的附加值,走好乡村绿色、可持续的发展之路。通过组织专家进行实地考察,确定了河北兴隆暗夜公园星空驿站项目和浙江凤溪玫瑰教育研学基地项目两个赛题,并编制了设计任务书。

"2019台达杯国际太阳能建筑设计竞赛"正式启动
The International Solar Building Design Competition 2019 was initiated

This competition is sponsored conjointly by the International Solar Energy Society, Chinese Renewable Energy Society (CRES) and National Supervision Board of Architectural Education (China), organized by China National Engineering Research Center for Human Settlements and the Special Committee of Solar Buildings, CRES, and co-organized by China Architecture Design & Research Group with the title sponsor of the Delta Group. With great support from all sectors of society, the Organizing Committee for this competition has organized a series of activities ranging from competition start up, media campaigns, campus tours, registration and submission and work evaluations, etc. These activities have received positive responses and active participation from industry experts at home and abroad.

1. Competition Preparation

The ways to develop rural cultural tourism and effective transformation from rural ecological resources into economic sources can boost industrial prosperity, create an ecologically livable environment, enhance rural civilization, form an effective mechanism for governance, thus ultimately achieving a prosperous life, and providing new opportunities for rural rejuvenation through cultural development. The competition, themed with "Sunshine & A Tour of Culture", aims to further activate various rural resources against the backdrop of rural revitalization, enhance the added value of rural development for a rural path featuring green and sustainable development through research bases, stations and other carriers. Through the previous field tips, two subjects were set up: Starry Sky Station in Xinglong Night Park Project, Hebei Province and Fengxi Rose Education Research Study Base Project, Zhejiang Province, and task books for designing of the two contest themes were compiled.

2. Competition Start up

On August 21, 2018, the International Solar Building Design Competition 2019 was initiated in Beijing. Many guests attended the opening ceremony of

参赛团队现场调研竞赛场地
Partcipating team research competition venue

二、竞赛启动

2018年8月21日，2019台达杯国际太阳能建筑设计竞赛在北京启动。中国可再生能源学会理事长谭天伟、中国建筑学会秘书长仲继寿、中国建筑设计研究院有限公司董事长宋源、台达品牌长郭珊珊等嘉宾出席并参加了竞赛启动仪式，共同为"2019台达杯国际太阳能建筑设计竞赛"的两个赛题揭幕。

本届竞赛在南北两个气候区分别设置赛题。其中，河北兴隆暗夜公园星空驿站项目要求在用地范围内，结合自然环境和星空主题，设计星空驿站，主要为暗夜公园的游客、天文台工作的科学家、学者等提供住宿、科普活动空间。浙江凤溪玫瑰教育研学基地将建设教学中心、创意中心、学生宿舍、体验单元等。赛题要求结合实际项目的建设需求，充分利用主动、被动太阳能技术，结合周边优越的自然环境，建设适用于寒冷地区、夏热冬冷地区的文化旅游设施。

三、校园巡讲

国际太阳能建筑设计竞赛巡讲是本项活动的重要组成部分，自启动以来，得到了清华大学、天津大学、东南大学、重庆大学、山东建筑大学等国内众多建筑

the competition, including Tan Tianwei, President of the CRES, Zhong Jishou, Secretary General of the Architectural Society of China, Song Yuan, Chairman of China Architecture Design & Research Group, and Guo Shanshan, Chief Brand Officer of Delta, and they launched the inauguration ceremony for the competition.

The competition set up the subject in two different climate zones in north and south of China. The Hebei Xinglong Night Park Starry Sky Station project requires the design of a starry sky station within the scope of land use, combined with the natural environment and the starry sky theme, mainly providing accommodation and popular science activities for tourists in the dark night park, scientists and scholars working in the observatory. Zhejiang Fengxi Rose Education Research Base will build teaching centers, creative centers, student dormitories and experience units, etc. The subject requires the use of actual project construction needs, make full use of the active and passive solar technology, combined with the surrounding natural environment, to build cultural and tourism facilities suitable for cold areas, and areas characterized by a hot summer and a cold winter.

3. Campus Tours

Campus Tours for the International Solar Building Design Competition constitute an integral part of this event. Since its initiation, Tsinghua University, Tianjin University, Southeast University, Chongqing University, Shandong Jianzhu University and many other domestic architectural colleges and universities have given us their support. As a result, the tours have gradually become public benefit activities with much influence on campuses and attracted a large number of passionate young designers to participate in the competition.

On October 24, 2018, the Organizing Committee of International Solar Building Design Competition went to Shanghai Institute of Technology, and exchanged their ideas with teachers and students there centered on the competition. The members of campus tours for the competition also left for several colleges and universities, such as Shanghai JiGuang Polytechnic College, China University of Mining and Technology, Beijing, Shihezi University

中国矿业大学（北京）巡讲现场
Scenes of campus tour in China University of Mining and Technology (Beijing)

济南大学巡讲现场
Scenes of campus tour in University of Jinan

重庆大学巡讲现场
Scenes of campus tour in Chongqing University

浙江理工大学巡讲现场
Scenes of campus tour in Zhejiang Sci-Tech University

院校的大力支持，逐渐成为一项具有影响力的校园公益活动，也吸引了大批富有激情与梦想的青年设计师积极参与竞赛。

2018年10月24日，国际太阳能建筑设计竞赛组委会走进上海应用技术学院，与师生们围绕国际太阳能建筑设计竞赛进行了交流。竞赛巡讲团一行还前往了上海济光学院、中国矿业大学（北京）、石河子大学和新疆大学等院校进行校园巡讲。随后，竞赛巡讲团继续前往广州、浙江等地，分别在广州大学、广东工业大学、浙江理工大学、浙江工业大学、浙江大学开展校园巡讲。巡讲主讲人为国家住宅与居住环境工程技术研究中心曾雁总建筑师，巡讲内容涵盖了太阳能建筑技术应用趋势和现状、历届竞赛获奖作品分析和本届竞赛介绍。通过巡讲，师生们对太阳能建筑设计竞赛和节能技术有了更深入的了解，激发了参赛团队的设计灵感，对太阳能建筑应用技术进行了创新思考。

四、媒体宣传

自竞赛启动伊始，组委会通过多渠道开展媒体宣传工作，包括：竞赛双语网站实时报道竞赛进展情况并开展太阳能建筑的科普宣传；在百度设置关键字搜索，方便大众查询，从而更快捷地登陆竞赛网站。在中国《建筑学报》、《建筑技艺》等专业杂志刊登了竞赛活动宣传专版，在新华网、腾讯网、新浪网等50余家网站上报道或链接了竞赛的相关信息；同时，组委会与多所国外院校取得联系并发布竞赛信息。

and Xinjiang University. Later, they continued to start their trips to some places including Guangzhou City and Zhejiang Province, and delivered campus tours respectively in Guangzhou University and Guangdong University of Technology, as well as Zhejiang Sci-Tech University, Zhejiang University of Technology and Zhejiang University. As the keynote speaker of campus tours, Zeng Yan, Chief Architect of the China National Engineering Research Center for Human Settlements, made a speech centered on the trend and status of applying solar building technologies, analysis of prize awarded works from previous competitions, and introduction to the competition of this year. The campus tours provided a sound opportunity for teachers and students to have an in-depth understanding of the solar building design competition and energy saving technology, stimulated aspirations for participant teams' design, and helped them trigger innovative thinking on the solar building application technology.

4. Media Campaign

Since the competition's initiation, the Organizing Committee has been making great efforts to popularize its media influence through many channels, including a bilingual website that reported real-time competition progress and spread scientific knowledge for solar buildings, set keyword searches on Baidu,

五、竞赛注册及提交情况

本次竞赛的注册时间为 2018 年 8 月 21 日至 2019 年 1 月 1 日，共 936 个团队通过竞赛官网进行了注册。其中，境外注册团队 10 个，包括日本、巴基斯坦、突尼斯、芬兰、加拿大、德国等国家和中国港澳台地区。截至 2018 年 3 月 2 日，竞赛组委会共收到参赛有效作品 200 份。

六、作品初评

2019 年 3 月 5 日，组委会将全部有效作品提交给初评专家组。每位专家根据竞赛办法中规定的评比标准对每一件作品进行评审，按照作品票数由高到低，共有 102 份作品进入中评。经过专家的严格审查，组委会对所有专家的评审结果进行统计之后，按照票数由高到低，共 60 份作品进入终评阶段。

enabling public searches much more convenient and much easier to log into the competition website, published special edition for competition advertising in China's *Architectural Journal, Architecture Technique* and other professional magazines, and delivered the report or set up links of relevant competition information on more than 50 websites, such as xinhuanet.com, qq.com and sina.com. Meanwhile, the Organizing Committee reached out to many foreign colleges and universities, and released the competition information.

5. Registration and Submission

The registration time of the competition in 2019 ranged from August 21, 2018 to January 1, 2019, and a total of 936 teams made registration via the competition website. Among them, there were 10 registered teams outbound of

竞赛官方网站和宣传报道　Oiffcial website and media reports

评委参观落成项目
Jury menbers visited the completed project

China, including such countries as Japan, Pakistan, Tunisia, Finland, Canada and Germany, as well as China's Hong Kong, Macao and Taiwan regions. As of March 2, 2018, the Organizing Committee had received 200 effective works.

6. Preliminary Evaluation

On March 5, 2019, the Organizing Committee submitted all the valid works to the jury of preliminary evaluation. Each expert reviewed all the works according to the evaluation requirements regulated in the Competition Evaluation Methods, and selected 102 works into the next evaluation process by an order of decreasing number of votes for works. After strict review of those experts, the Organizing Committee delivered the evaluation statistics from all the experts, and selected 60 works for the final evaluation in accordance with the vote orders from high to low.

7. Final Evaluation

The final evaluation conference was conducted in Tongli, Jiangsu on April 24, 2019. Through the discussion of expert groups at the conference, Professor Cui Kai was unanimously selected to assume the group leader of the work evaluation in this competition. During his presiding over the conference, the evaluation jury made collective discussion and fair evaluation about the works according to the principle of simple majority. Three-round votes saw 42 award works selected, two of which won the first prize, four of which the second prize, six of which the third prize, and 30 of which the excellence award.

七、作品终评

竞赛终评会于2019年4月24日在江苏同里召开。在终评会上，经专家组讨论，一致推选崔愷院士担任本次终评工作的评审组长。在他的主持下，评审专家组按照简单多数的原则，集体讨论和公正客观地评选作品，通过三轮的投票，共评选出42项获奖作品，其中一等奖2名、二等奖4名、三等奖6名、优秀奖30名。

终评会现场　Scenes of final evaluation conference　　终评专家组合影　Members of final evaluation juries

2019台达杯国际太阳能建筑设计竞赛评审专家介绍
Introduction of Jury Members of International Solar Building Design Competition 2019

评审专家
Jury Members

杨经文：马来西亚汉沙杨建筑师事务所创始人、2016梁思成建筑奖获得者
Kenneth King Mun YEANG, President of T. R. Hamzah & Yeang Sdn. Bhd. (Malaysia), 2016 Liang Sicheng Architecture Award Winner

Deo Prasad：澳大利亚科技与工程院院士、澳大利亚勋章获得者、新南威尔士大学教授
Deo Prasad, Professor of University of New South Wales, Sydney, Australia, Asia-Pacific President of International Solar Energy Society (ISES) and Professor of Faculty of the Built Environment, the Order of Australia

林宪德：台湾绿色建筑委员会主席、台湾成功大学建筑系教授
Lin Xiande, Chairman of the Taiwan Green Building Committee, Professor of Cheng Kung University Tai Wan

崔愷：中国工程院院士、全国工程勘察设计大师、中国建筑设计研究院有限公司总建筑师
Cui Kai, Academician of China Academy of Engineering, National Design Master and Chief Architect of China Architecture Design & Research Group

王建国：中国工程院院士、全国高等学校建筑学学科专业教学指导委员会主任
Wang Jianguo, Academician of China Academy of Engineering, Director of National Supervision Board of Architectural Education (China)

仲继寿：中国建筑学会秘书长、中国可再生能源学会太阳能建筑专业委员会主任委员
Zhong Jishou, Secretary general of the Architectural Society of China; Chief Commissioner of Special Committee of Solar Building, CRES

庄惟敏：全国工程勘察设计大师、清华大学建筑学院院长
Zhuang Weimin, National Design Master; Dean of School of Architecture, Tsinghua University

黄秋平：华东建筑设计研究总院总建筑师
Huang Qiuping, Chief Architect of East China Architecture Design & Research Institute

冯雅：中国建筑西南设计研究院有限公司副总工程师、中国建筑学会建筑热工与节能专业委员会副主任
Feng Ya, Deputy Chief Engineer of China Southwest Architectural Design and Research Institute Corp., Ltd; Deputy Director of Special Committee of Building Thermal and Energy Efficiency, Architectural Society of China

获奖作品
Prize Awarded Works

风吟·花舞 1 2019台达杯国际太阳能建筑设计竞赛
WIND SINGING · FLOWERS DANCING
2019 Taida Cup International Solar Building Design Competition

综合奖·一等奖
General Prize Awarded · First Prize

注 册 号：6122
项目名称：风吟·花舞（凤溪）
　　　　　Wind Singing · Flowers Dancing (Fengxi)
作　　者：谢瑞航、黄圣翔、叶珍光、马金辉
参赛单位：重庆大学
指导老师：周铁军、张海滨

专家点评：

该作品建筑平面规划与场地条件结合较好，建筑功能和形态适宜南方湿热潮气候。利用太阳能相变蓄热、通风烟囱调节室内通风，在夏季及过渡期有良好的自然通风及热交换（烟囱效应），蓄热墙与中庭采光通风有效调节冬夏两季室内热环境。冬季蓄热墙有效地保证室内温度，还可按时序进行蓄放热。组团建筑形态和主动太阳能技术方面需要进一步提高。

The works combines the architectural plan with the site condition well, and the architectural function and shape are suitable for hot and wet climate in south of China. With the solar collectors and LHTS, as well as the air chimney to adjust indoor ventilation, a sound natural ventilation and heat exchange (chimney effect) state can be achieved in summer and the transition period, and the lighting and ventilation between the thermal storage wall and atrium can effectively regulate the indoor thermal environment in winter and summer. The lighting and ventilation in winter can effectively sustain the indoor temperature, and store and reject heat according to time series. However, both the architectural pattern of groups and active solar technology should be further improved.

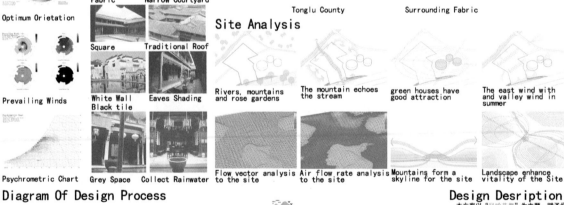

Diagram Of Design Process

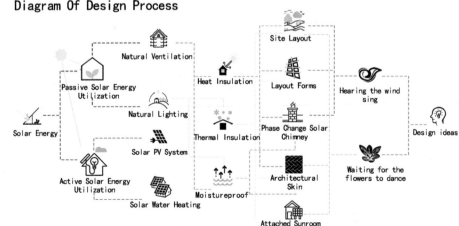

Design Desription

本方案以"风吟花舞"为主题，赋予场地自然的物质与情感属性。

因此，本方案以被动式技术为设计出发点，结合人群活动行为，桐庐传统民居形式，重点加强建筑过渡季节与夏季自然通风处理，兼顾冬季保温。通过场地布局和设置热缓冲空间，利用庭院、灰空间、太阳能相变蓄热烟囱等来实现建筑节能。同时，在主动式技术上，根据建筑使用的不同，分别设置太阳能光伏系统和太阳能光热系统，并于建筑屋顶一体化设计。此外加入了雨水收集系统，将收集的雨水用于景观环境和冲厕等非生活用水。

The theme of this project is "Wind Singing Flower Dance", which endows the site with natural material and emotional attributes.

Therefore, this scheme takes passive technology as the design starting point, combined crowd activity behavior, Tonglu traditional residential form, focusing on strengthening natural ventilation treatment in summer, taking into account the winter heat preservation. Through the layout of the site and the setting of thermal buffer space, the use of courtyards, gray space, solar phase change chimney and so on to achieve building energy saving. At the same time, in the active technology, we set up solar PV system and solar PT system according to the different building requirement, we add a rainwater collection system to for non-domestic water such as landscape environment and toilet flushing.

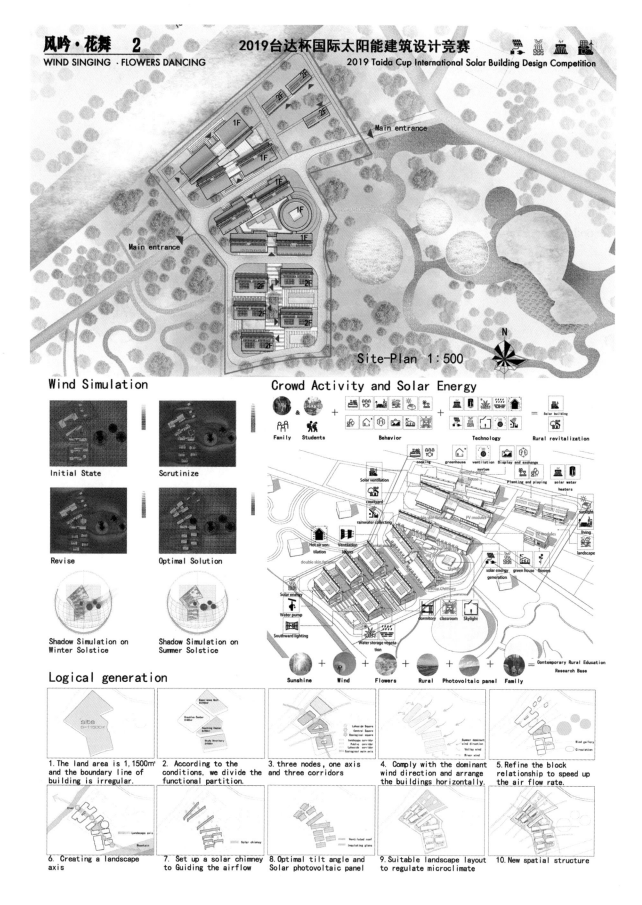

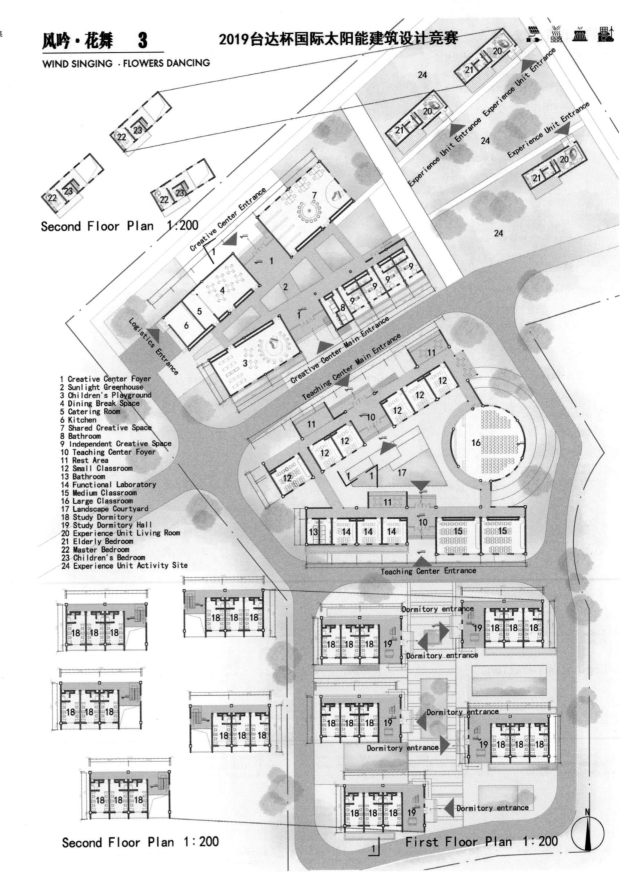

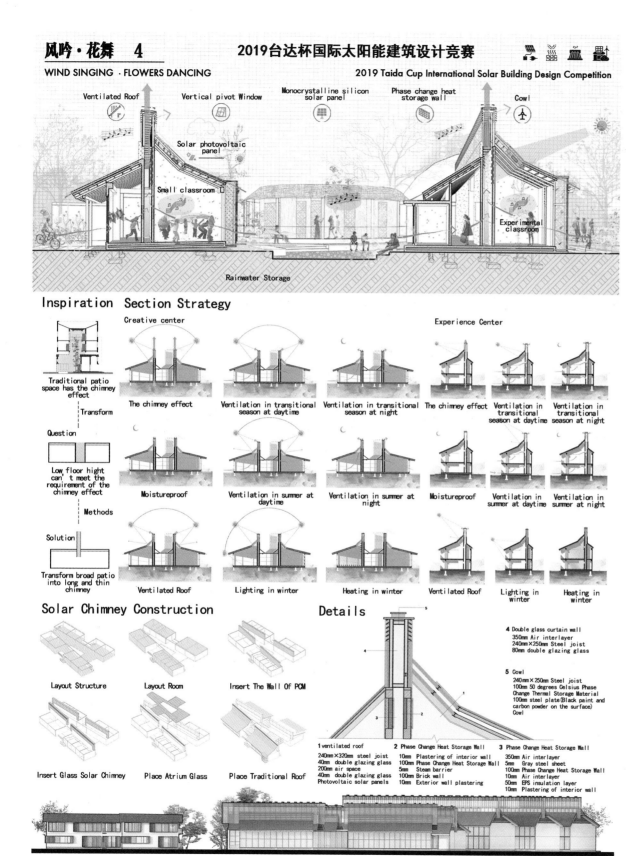

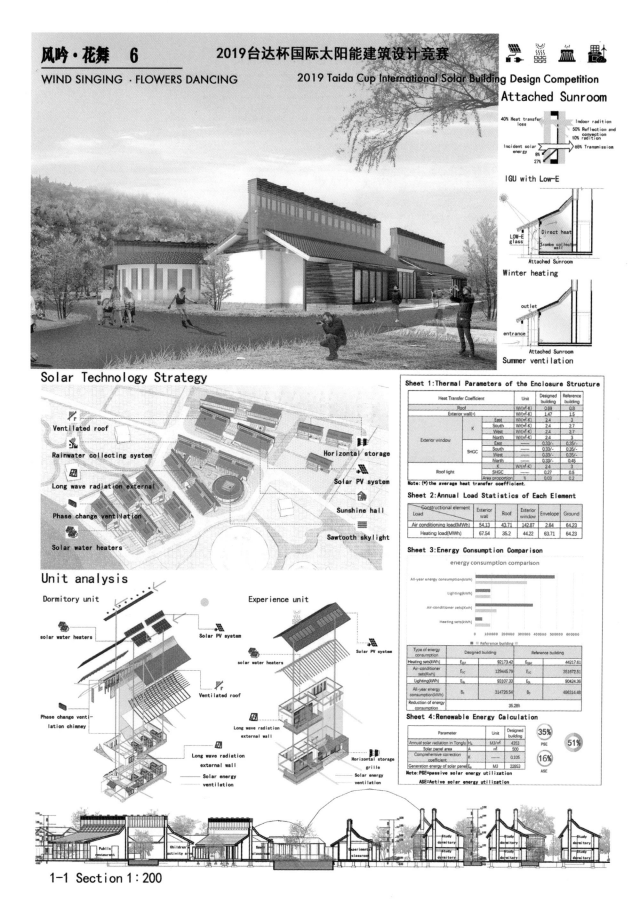

综合奖·一等奖
General Prize Awarded · First Prize

注 册 号：6561
项目名称：摘·星辰（兴隆）
　　　　　Pick Stars（Xinglong）
作　　者：戴芙蓉、徐　涵、陈晓能
参赛单位：南京工业大学
指导老师：杨亦陵

专家点评：

该作品构思巧妙，设计得体有特色，充分考虑用地特征，结合山地地貌进行布局。单体建筑组合成的三合院组团通过高低错落的灵活栈道与环境高度融合，建筑群错落有致。布局考虑谷底的架空、空间的通透，契合了环境与气候。建筑单体设计充分考虑了自然通风、太阳能光伏板、冬夏季遮阳转换及雨水收集的被动式技术与建筑设计相结合。作品中对被动式技术的运用深度不足。

The work, ingenious in thinking and characteristic in design, conducts the overall arrangement through full consideration of land characteristics and layout of mountainous landform. The triple dwelling group formed by single buildings highly integrates scattered high and low plank roads with the environment, so the building clusters are well-proportioned. The layout fits the environment and climate by considering the overhead bottom of the alley and space transparency. The single building's design takes into account natural ventilation, solar photovoltaic panels, sunshade conversion in winter and summer, as well as the combination of passive technology for rainwater collection and architectural design. However, the use of passive technology in the work is insufficient.

摘·星辰 STAR STATION IN NIGHT PARK 1

DESIGN NOTES This design is located in Xinglong Night Park, Hebei province. Star station pursues supreme star-watching experience. In the process of design, the combination of star-watching places, mountainous conditions and the concept of green construction is the divergence of design thinking. The design feature is the ingenious combination of stargazing sites and green construction facilities. Extraction of the triple courtyard form in Hebei province and the structures with cohesive function. Each group is scattered like a constellation in the field. The buildings on both sides of the valley rae collected by corridors. Travelling under the stars, it seems like that we can pick the stars.

设计说明 本次设计位于河北兴隆暗夜公园项目内，暗夜公园是国际暗夜协会为呼吁治理全球光污染而在全球范围内评选的一些暗夜条件最好的公园。因此，星空驿站首要追求极致的观星体验。在设计过程中观星场所、山地条件以及绿建理念的结合是思维发散点。本设计的特色正是体现在观星场所与绿建设施巧妙地结合。提取自河北当地民居的三合院形式结合具有凝聚作用的构筑物，每一组如星群般散落在场地内。山谷两侧建筑以连廊相接，星空下穿行其间，人们恍若置身满船清梦压星河场景中。

ECONOMIC AND TECHNICAL INDICATERS

Land area 7868㎡	Total building area 2400㎡	Building occupation area 1760㎡
Building density 22%	Plot ratio 0.3	The height of public building 13m
The height of guest rooms 4m		

经济技术指标

用地面积 7868㎡	总建筑面积 2400㎡	建筑占地面积 1760㎡
建筑密度 22%	容积率 0.3	公建部分高度 13m
客房部分高度 4m		

Site analysis

Site line

Orientation
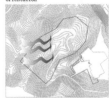

Surrounding building
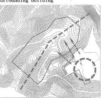

Functional zoning

Site reality

The vegetation in the field is flourishing. The terrain fluctuates greatly.

摘·星辰 STAR STATION IN NIGHT PARK 2

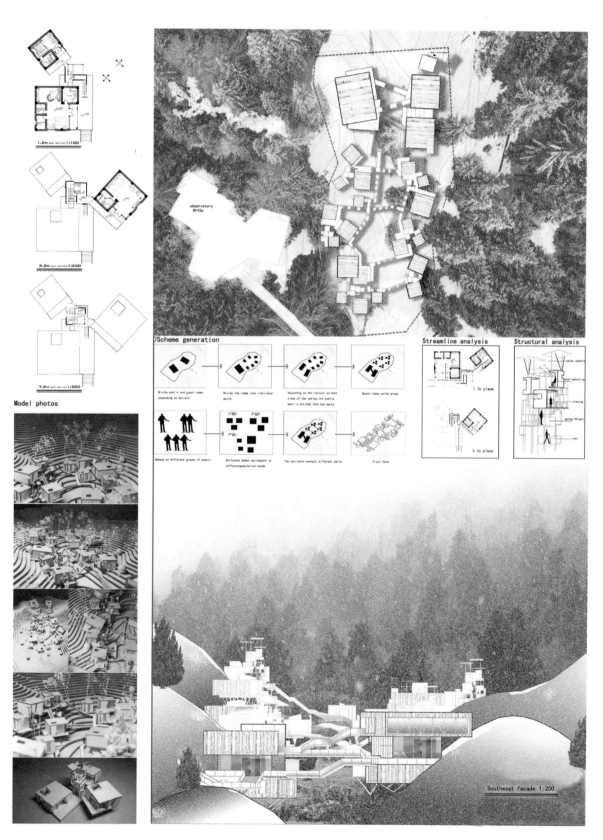

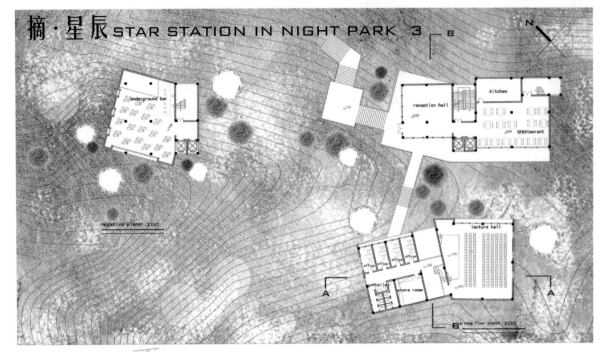

摘·星辰 STAR STATION IN NIGHT PARK · 4

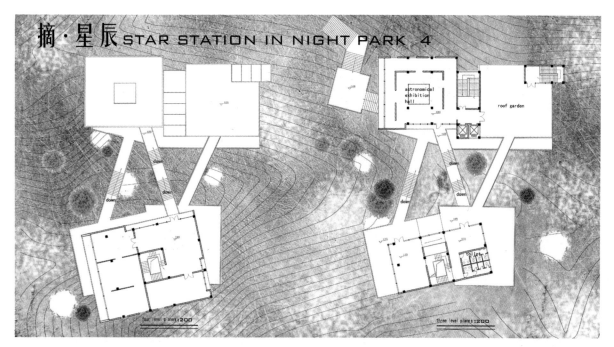

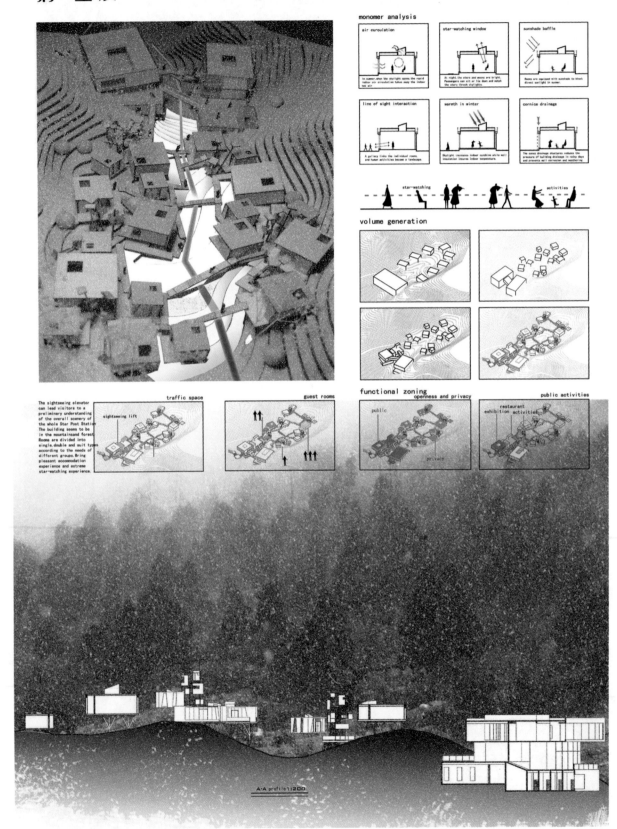

摘·星辰 STAR STATION IN NIGHT PARK 6

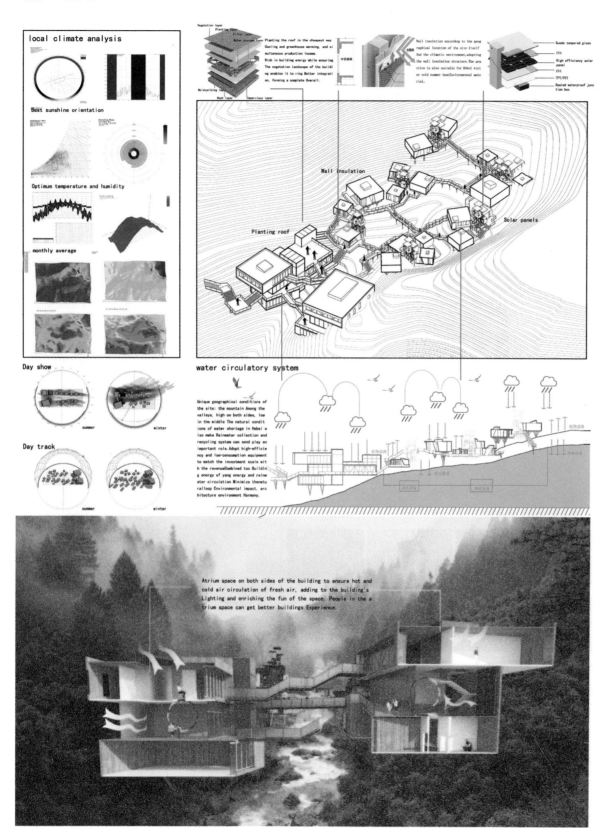

综合奖·二等奖
General Prize Awarded · Second Prize

注　册　号：6522
项目名称：溪·聚（凤溪）
　　　　　　Gathered by River（Fengxi）
作　　　者：苏　红、陈　剑、冯温然、
　　　　　　沈伟斌、吴锦凯、王嘉琪、
　　　　　　颜君家
参赛单位：浙江理工大学
指导老师：文　强

专家点评：

该作品以"溪·聚"作为规划和建筑布局理念。教学区、创意中心、宿舍和体验单元通过连续廊道和水溪组织，总体布局较灵动活泼，符合亲子教育活动的特点和要求。建筑设计具备系统性，对朝向、自然通风、院落等被动节能因素考虑较为充分，同时也对不同季节的太阳能、光伏板及风、光、热等因素进行了综合考虑。技术集成性良好。在场地设计中，取消了原场地东西向道路，未充分表达其取消理由。建筑造型尚待改进，水面面积较多也欠妥。

The work takes "Gathered by River" as the planning and architectural layout concept. The teaching area, creative center, dormitory and experience unit pass through continuous corridors and waterways, so the overall layout is more dynamic, which conforms to the characteristics and requirements of parent-child education activities. The building design is systematic, in which the author considers the passive energy-saving factors such as orientation, natural ventilation and courtyard, and also fully considers the solar energy, photovoltaic panels, wind, light and heat in different

设计说明
以水为带，串络泱泱绿意；
以情为系，共赴欢聚之堂。

设计源于桐庐古镇的特色水圳，以梳理场地排水系统来组织不同层次的聚合空间。被动式设计上，应用建筑朝向、阳光房、院落、风塔、自然通风、热压通风等方式，并采用了太阳能集热板、光伏电板、地源热泵、沼气等主动式设计，实现了对光、风、水、沼气、地热等可再生资源的利用。设计在自然生态环境中以溪水串联教学区、创意中心、宿舍区、体验单元四大空间，营造各种层次的交流共聚场所。

DESIGN DESCRIPTION
The design originates from the characteristic water system of Tonglu Ancient Town, which organizes the drainage system of the site to organize different levels of polymerization space. Passive design, application of building orientation, sun space, courtyard, wind tower, natural ventilation, hot pressure ventilation, etc. and the use of solar collector panels, photovoltaic panels, ground source heat pumps, biogas and other active design, achieved the use of renewable resources such as light, wind, water, biogas, and geothermal. Designed in the natural ecological environment, the river connects teaching area, the creative center, the dormitory area, and the experience unit—four spaces to create various levels of communication and copolymerization places.

CLIMATE ANALYSIS

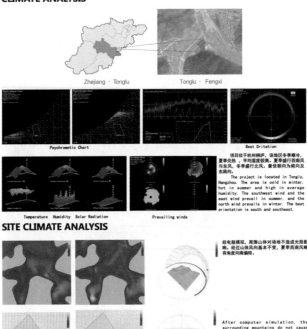

项目位于杭州桐庐，该地区冬季寒冷，夏季炎热，平均湿度较高。夏季盛行西南风与东风，冬季盛行北风。最佳朝向为南向和东南向。

The project is located in Tonglu, Hangzhou. The area is cold in winter, hot in summer and high in average humidity. The southwest wind and the east wind prevail in summer, and the north wind prevails in winter. The best orientation is south and southeast.

SITE CLIMATE ANALYSIS

经电脑模拟，周围山体对场地不造成光照影响。经过山体风向基本不变，夏季西南风略有角度向南偏转。

After computer simulation, the surrounding mountains do not cause light effects on the site. After the mountain wind direction is basically unchanged, the southwest wind in summer is slightly angled to the south.

TRADITIONAL ELEMENTS ANALYSIS

Drainage System
水圳 WATER CANAL

小天井 Small Patio

Enhance Ventilation
小天井 Small Patio　　　复合式天井 Combined Patios

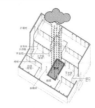 

Thermal Insulation
卵石 Local Stone

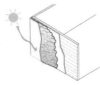 

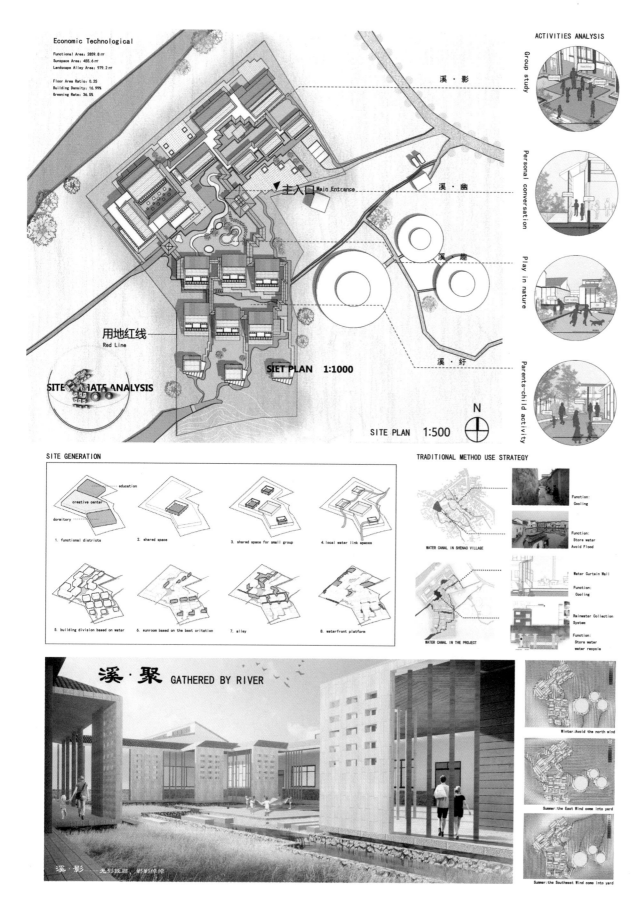

seasons. The technology is well integrated. In the site design, the east-west road of the original site is cancelled, but the reason is not fully expressed. The architectural shape still needs to be improved, and considerations about the large area of water surface is inadequate as well.

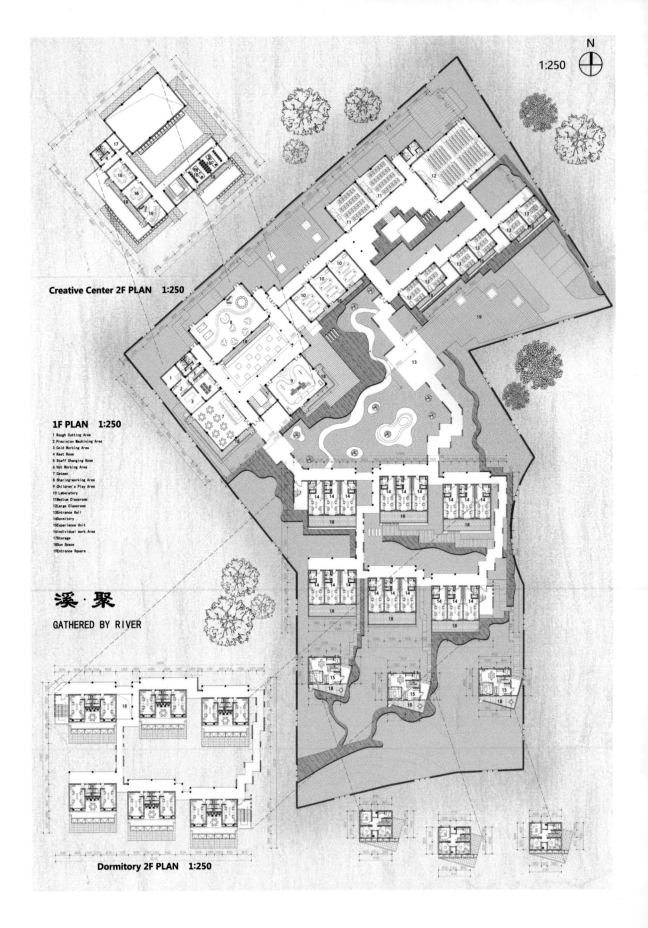

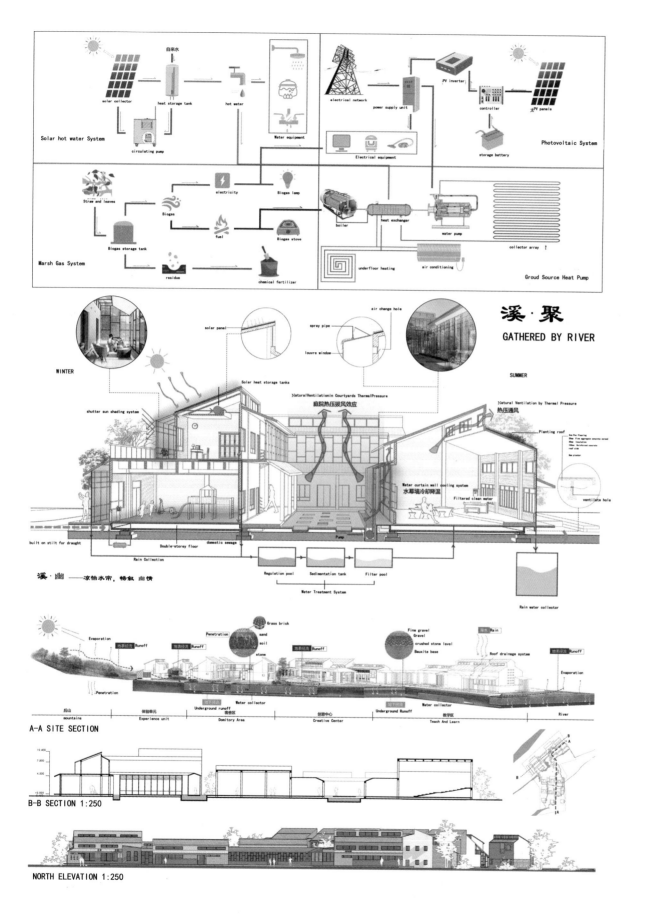

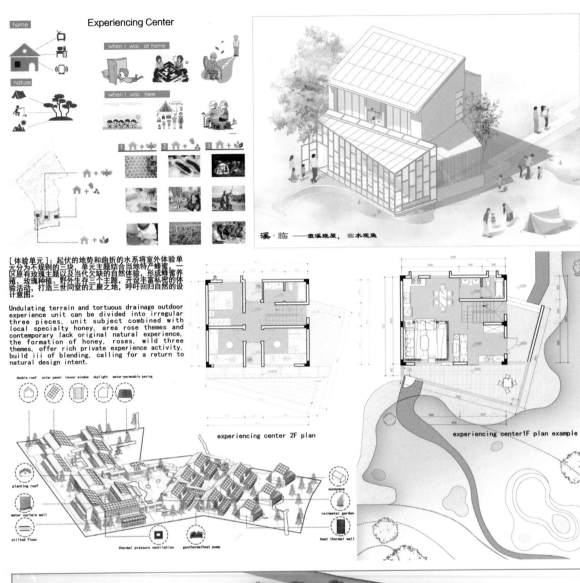

[体验单元]：起伏的地势和曲折的水系将室外体验单元分为不规则的三块，单元主题结合当地特产蜂蜜，一区原有玫瑰主题以及当代欠缺的自然体验，形成蜂蜜养殖、玫瑰种植、野外生存三个主题，开设丰富私密的体验活动，打造三世同堂的汇聚之地，呼吁回归自然的设计意图。

Undulating terrain and tortuous drainage outdoor experience unit can be divided into irregular three pieces, unit subject combined with local specialty honey, area rose themes and contemporary lack original natural experience, the formation of honey, roses, wild three themes, offer rich private experience activity, build iii of blending, calling for a return to natural design intent.

溪·聚 GATHERED BY RIVER

研学宿舍单元分解图
Study Dormitory Unit

Sun Blinds 遮阳百叶
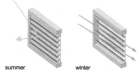
summer / winter

Thermal Mass Wall 蓄热墙体
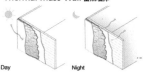
Day / Night

Planting Roof 屋顶绿化
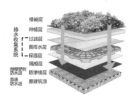

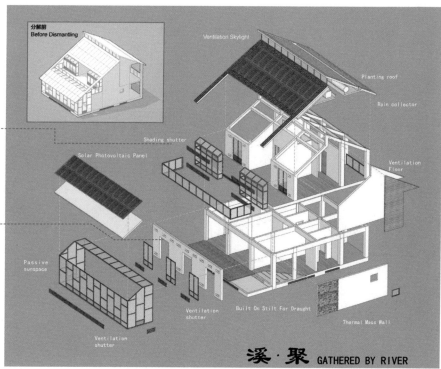

溪·聚 GATHERED BY RIVER

Passive Sunroom 被动式阳光房
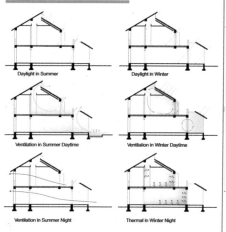

The First Floor | The Second Floor

溪·纡 ——溪泉碧溶，纡径观景

综合奖·二等奖
General Prize Awarded · Second Prize

注 册 号：5822
项目名称：倚山居（兴隆）
　　　　　Lean on the Mountain
　　　　　（Xinglong）
作　　者：王子嘉、谢颂、苏晓婉
参赛单位：北京交通大学
指导老师：杜晓辉、胡映东

专家点评：

本作品建筑均设置在北坡，房间面对南坡，充分利用太阳采暖，符合当地气候特点。建筑配置集约，节省用地，外形表面积小，保有节能体形系数。阶梯型布置与坡地自然结合，南山坡绿地完全被保留，充分将景观与生态环境进行融合。作品在太阳光伏、阳光房等方面技术运用不足。

The buildings in this work are all set on the north slope, and the room is oriented towards the south one. They make full use of the sun for heating, which is in compliance with the local climate characteristics. The building is intensively constructed, which saves land with a small surface area and an energy-saving shape factor. The ladder type is naturally combined with the slope land, and the green land of the south slope is completely preserved, fully integrating the landscape with ecological environment. However, technologies such as solar photovoltaics and sun room are not applied efficiently in the work.

倚山居 Lean on the mountain 2

In order to get enough sunshine and a continuous and open view, we have arranged all the units on the south hillside and adopted retreat-platform style to increase the Interest.

Layout generation

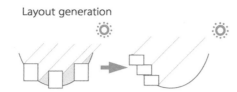

General Layout 1:500

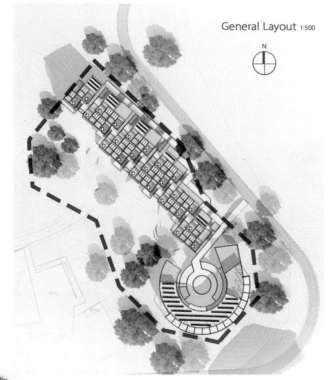

Sight analysis

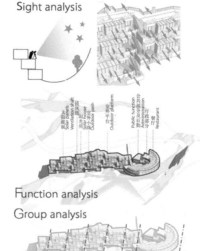

Function analysis
Group analysis
Traffic analysis

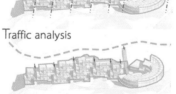

倚山居 Lean on the mountain

注册号5822

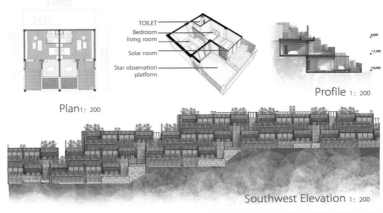

Plan 1:200
Profile 1:200

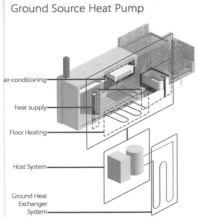

Ground Source Heat Pump

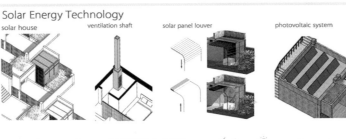

Southwest Elevation 1:200

Solar Energy Technology
solar house | ventilation shaft | solar panel louver | photovoltaic system

Reducing surface runoff (Sponge City)
- Seepage brick
- roof planting
- Overhead wood flooring

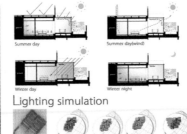

Summer day | Summer day(wind) | Winter day | Winter night

Lighting simulation

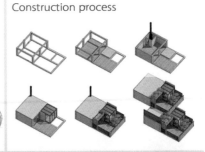

Construction process

倚山居 Lean on the mountain 4

2019 台达杯国际太阳能建筑设计竞赛获奖作品集

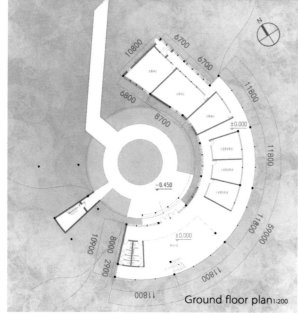

Ground floor plan 1:200

Second floor plan 1:200

"round sky and square earth"

Third floor plan 1:200

The north side of the building is provided with continuous strip windows at the human visual height, so that tourists can see the valley scenery in the building.

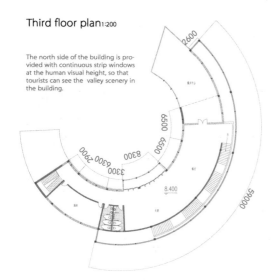

Solar house

The public building is located in the valley of the mountain. According to the shape of the mountain, each floor is part of ring, with three layers staggered to reduce the sense of consideration.

photovoltaic system

West elevation 1:200

Profile 1:200

综合奖·二等奖
General Prize Awarded · Second Prize

注 册 号：6041
项目名称：通往星空——河北兴隆暗夜公园星空驿站设计（兴隆）
The Way to the Starry Sky—Design of Starry Sky in Xinglong Dark Night Park in Hebei（Xinglong）
作　　者：张潇方、邱丛丛、陈家乐
参赛单位：武汉大学
指导老师：黄凌江、李鹍

专家点评：
该作品针对暗夜公园设施的独特功能和环境，以独立单元点状分布的布局策略减少对山地的干扰。采用天窗营造独特的观星环境。在建筑造型上兼顾朝向和防风雪的要求，并由此形成富有趣味的空间形态，与山地景观呼应。但该作品在太阳能主动利用和被动式节能技术应用的表达不够系统化，场地中间栈道的规划设计欠妥，观星天窗视野过小，需要优化和改进。

With respect to the unique functions and environment of the dark night park facilities, the work aims to reduce the interference to mountains through the layout strategy of a dotted pattern of independent units. Skylights are used to create a unique star-gazing environment. The architectural shape takes into account the requirements of orientation as well as wind and snow prevention, thus forming an interesting spatial form that echoes the mountain landscape. However, the expression of the active utilization of solar energy and the application of passive energy-saving technology are not

本方案以"通往星空"为主题，旨在为暗夜公园留宿的游客提供一种穿梭于林间缓缓通向星空的特别住宿体验。一个个伸向天空的太阳能烟囱客房，像是人们望向星空的望远镜，勾起人们对宇宙的无限遐想。架起的漫游廊道、拥有开敞落地玻璃窗的客房、挑出的室外露台、透明的共享空间为观星提供了不同方式。同时本方案通过建筑自身的形态结合太阳能、利用雨水收集、采集当地材料，为游客提供一系列绿色舒适的观星活动空间。

With the theme of "the way to the starry sky", this plan aims to provide a special accommodation experience for the visitors who stay in the dark night park, which is to lead to the starry sky slowly through the forest. Each stretch to the sky of the solar chimney rooms, like people looking at the sky telescope, evoke people's infinite reverie of the universe. Roving corridors, guest rooms with open floor-to-ceiling Windows, outdoor terraces and transparent Shared Spaces provide different ways of stargazing. At the same time, this project provides tourists with a series of green and comfortable stargazing activity space by combining the form of the building with solar energy, rainwater collection and local materials collection.

总平面图/Total floor plan 1:500

气候分析/Climate Analysis

Xinglong county belongs to the warm temperate temperate semi - humid monsoon continental climate, four seasons obvious, winter long summer short. Annual average temperature between 6.5-10.3°. xinglong XianJing mountainous, vertical temperature change is obvious.

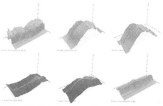

The winter is long and cold and dry, and the summer is hot and humid. The annual precipitation is 527mm, belonging to the region with less precipitation. The total solar radiation 507 kj/c·m², including direct radiation 299 kj/c·m², belong to the solar energy is relatively rich areas in China.

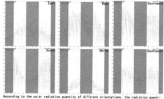

According to the solar radiation quantity of different orientations, the radiation quantity of south, southwest and west is higher, and fluctuates greatly in autumn and winter.

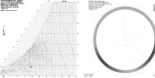

By enthalpy wet figure and best towards analysis, natural ventilation is passive technologies used in the field is more suitable and building toward appropriate south by east 20°.

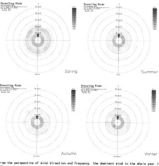

From the perspective of wind direction and frequency, the dominant wind in the whole year is NNW with an average wind speed of 2.5m/s. The dominant wind in summer is southerly(S), the average wind speed is 2.2m/ S; The dominant wind in winter is west-northwest (WNW) with an average wind speed of 2.5m/s.

systematic enough. Besides, the planning and design of the middle plank road is improper, and the star-gazing view through the skylight is so limited that it needs to be optimized and improved.

结论与策略/Conclusion

Best toward: south by east 20° and most of the solar radiation is minimal.
Enthalpy chart: the most comfortable periods are spring and autumn.
Solar radiation: the solar radiation of the north elevation is relatively small.
Natural ventilation: higher than 32° good natural ventilation can't solve the problem of cooling, you need to use air conditioning and mechanical ventilation.
Direct evaporative cooling: when the temperature is not higher than 33° can better solve the cooling problem.
Insulation of enclosure: the better the enclosure is, the higher the utilization of solar energy will be. Trumpwall can be used.
Passive solar heating: the south window rate of 40% can make solar heating to a better situation.
Active technology: air conditioning, mechanical ventilation and mechanical evaporation can better solve the problem of cooling.

场地区位分析/Site location analysis

SITE PHOTOS

1.Site area:7931.62m².

2.Function zoning and people flow line.

3.Our strategy is to avoid large areas of natural damage to the mountain.

4.Our strategy is to use pillars to set up a distributed volume and integrate with nature.

5.The building is arranged along the highest contour line of the mountain in the site, with the best stargazing perspective.

6.Large volumes of public and administrative areas are embedded in the mountain and disappear into nature.

7.Two or three guest rooms have a single Shared space, forming a number of groups. Zigzag ramps are used to shorten the streamline.

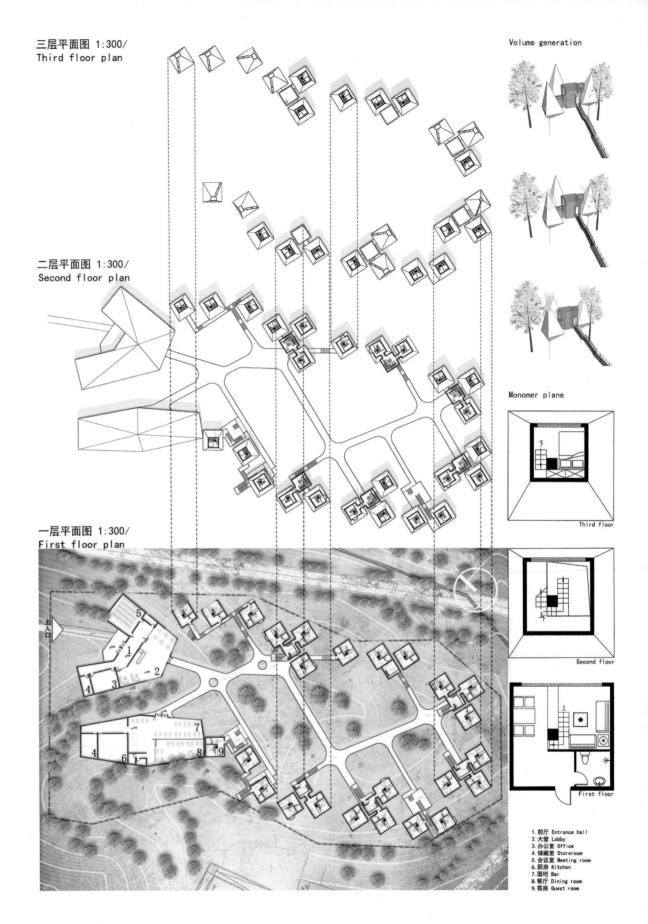

功能分析/Functional analysis

结构分析/Structure analysis

西立面图/West elevation 1:500

植物墙设计 / plant wall design

植物柱设计 / Plant column design

- Skeleton
- Growing Media 100% Recycled
- Hard-top (Recycled) Insulation Board
- Kempershield Vapour Barrier
- Thermal insulation layer
- Substrate Conservation

夏季通风-太阳能烟囱 / Ventilate in summer-Chimney Effect

冬季采暖-阳光房 / Thermal storage in winter-Sun Room

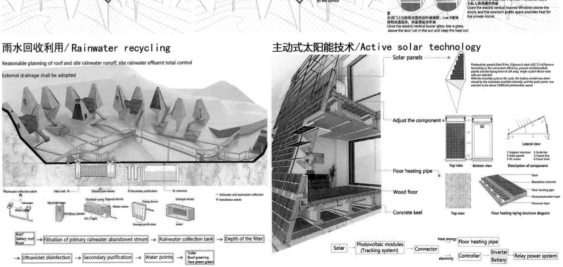

雨水回收利用 / Rainwater recycling

Reasonable planning of roof and site rainwater runoff, site rainwater effluent total control
External drainage shall be adopted

主动式太阳能技术 / Active solar technology

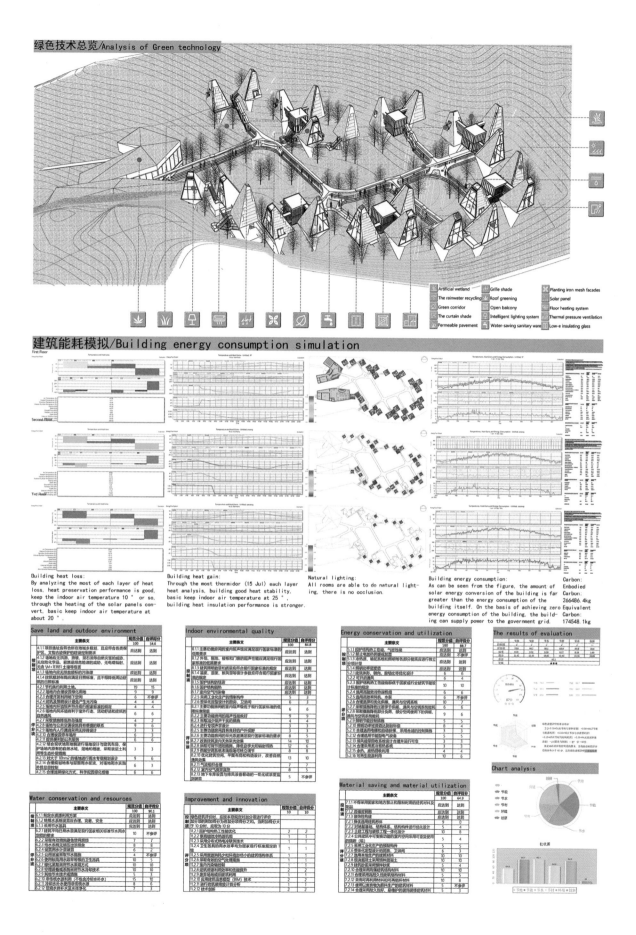

综合奖·二等奖
General Prize Awarded · Second Prize

注 册 号：6360
项目名称：云汉·七居（兴隆）
　　　　　The Galaxy & Dwellings
　　　　　（Xinglong）
作　　者：武浩然、巩振华、吴　昊、
　　　　　梁英伟
参赛单位：华南理工大学
指导老师：肖毅强、王　静

专家点评：

该作品采用扇形组团式布局，沿山路两侧等高线布置，扇形与圆形星桥相协调，总平面具有分散和集中相结合的优点。建筑体形系数小，组团建筑底层架空，尽可能减少对原始地形地貌的影响。每个建筑单元均为标准化设计，拥有室内、阳台和院落，空间丰富，有助于观星体验和装配式建造。太阳能光热技术运用适当。但方案总平面未充分考虑圆形观星桥对客房单元的影响，方案群体之间间距偏小，显得紧张。建议方案进一步加强室内外空间的丰富性，增加观星感受和整体观星小镇氛围。

The work adopts a fan-shaped group layout, arranged along the contour lines on both sides of the mountain road. The fan shape is coordinated with the circular star bridge, so the general plane features advantages of combining dispersion with concentration. The building shape coefficient is small, and the bottom of the group building is overhead, so as to minimize the impact on the original topography. Each building unit is designed by standards, with enormous space of interiors, balconies and courtyards, which is helpful to the star-gazing experience and

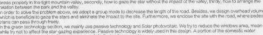
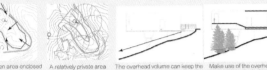

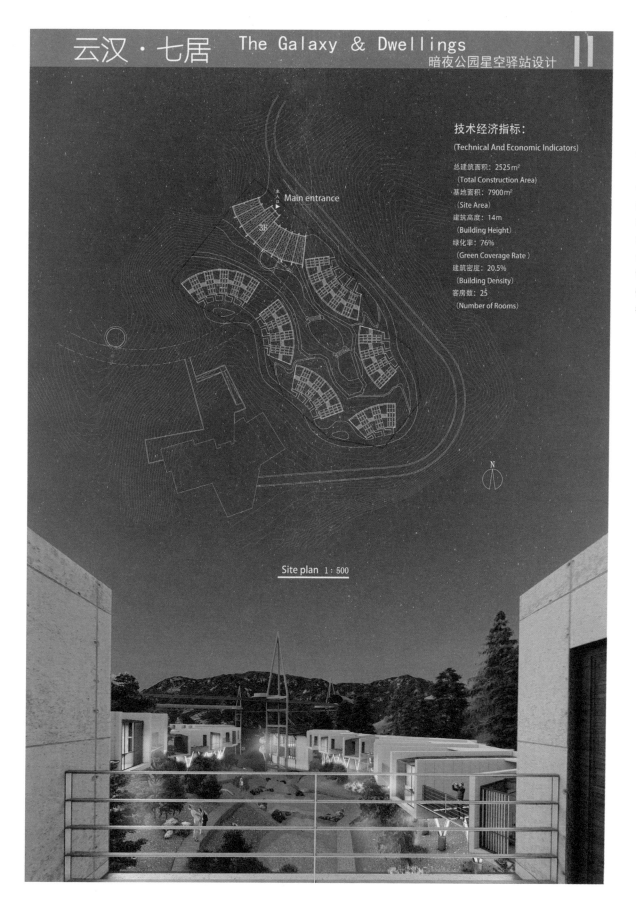

prefabricated construction. The utilization of solar thermal technology is appropriate. However, the overall plane doesn't fully consider the influence of the circular star-gazing bridge on the gest unit, and the space between the scheme groups is small and limited. It is suggested that abundance of interior and exterior space should be further improved, so as to enhance the star-gazing experience and atmosphere of the star-gazing town.

折院 ZIGZAG YARD
浙江凤溪玫瑰教育研学基地设计

01

综合奖·三等奖
General Prize Awarded · Third Prize

注 册 号：6249
项目名称：折院（凤溪）
　　　　　Zigzag Yard（Fengxi）
作　　者：张 琼、王飞雪、张吉强、
　　　　　杨 洋
参赛单位：天津大学
指导老师：朱 丽、孙 勇

专家点评：

该作品采用条形布局，功能合理，但对场地的切割略显生硬，各主要建筑间通过廊道连接，丰富了室外空间，建筑立面造型吸纳了当地民居特色，建筑室内空间较为丰富。遮阳、采光、通风措施与建筑结合较好。在主动式太阳能与建筑结合方面创新不足。

The work adopts a strip layout with reasonable functions, but the ways of site cutting is slightly blunt. The main buildings connected by corridors enrich the outdoor space. And the building facade takes in characteristics of local dwellings, with rather abundant indoor space. Also, the shading, lighting and ventilation measures are better combined with the building. However, there exists insufficient innovation in the combination of active solar energy and buildings.

SITE PLAN 1:500

ECONMIC INDICATORS
LAND AREA: 11589m²
BUILDING AREA: 2925m²
FLOOR AREA RATIO: 25%
GREENING RATE: 45%

DESIGN DESCRIPTION

建筑采用条形布局，以山为景象，曲折匍匐于田野之上，以求最大程度向自然打开，吸纳阳光和自然风。同时建筑根据不规则的场地围合出不同尺度院落空间，完成了对传统村落空间的抽象模拟。

建筑以竹、木等乡土材料为主，实现一种低技的绿色设计。设计立足于夏热冬冷气候区，从营造良好的风环境出发，综合利用多种被动式技术，同时采用了近零能耗热激活墙体耦合跨季节储能直接供能系统，达到更好的舒适和节能效果。

The building adopts a strip layout and lies prostrate at the open country, which uses local mountains as its background. These designs makes the building open to the environment, and therefore possessing a good ability of absorbing sunlight and utilizing wind to the greatest extent. At the same time, the building encloses courtyard space of different scales according to irregular site, and completes the abstract simulation for traditional village space.

Buildings are mainly made of bamboo, wood and other local materials to achieve a low-tech green design. The project is located in hot summer and cold winter zone. In order to create a better wind environment, a variety of passive technologies are synthetically utilized. At the same time, nearly-zero energy directing heating and cooling system is adopted to realize comfort and energy saving.

CLIMATE ANALYSIS

OPTIMIZATION OF MIEN VENTILATION

SITE

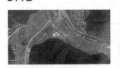

SITE MICROCLIMATE

FIRST FLOOR PLAN 1:300

BLOCK GENERATION

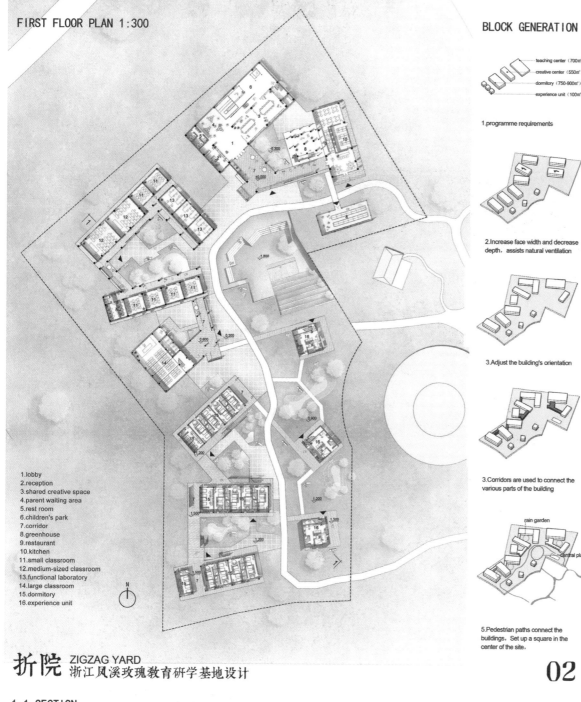

1. lobby
2. reception
3. shared creative space
4. parent waiting area
5. rest room
6. children's park
7. corridor
8. greenhouse
9. restaurant
10. kitchen
11. small classroom
12. medium-sized classroom
13. functional laboratory
14. large classroom
15. dormitory
16. experience unit

- teaching center (700㎡)
- creative center (550㎡)
- dormitory (750-900㎡)
- experience unit (100㎡*3)

1. programme requirements

2. Increase face width and decrease depth, assists natural ventilation

3. Adjust the building's orientation

3. Corridors are used to connect the various parts of the building

rain garden / Central plaza

5. Pedestrian paths connect the buildings, Set up a square in the center of the site.

折院 ZIGZAG YARD
浙江凤溪玫瑰教育研学基地设计

02

1-1 SECTION

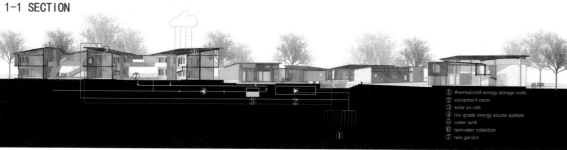

① thermal/cold energy storage wells
② equipment room
③ solar pv cell
④ low grade energy source system
⑤ water tank
⑥ rainwater collection
⑦ rain garden

折院 ZIGZAG YARD
浙江凤溪玫瑰教育研学基地设计

04

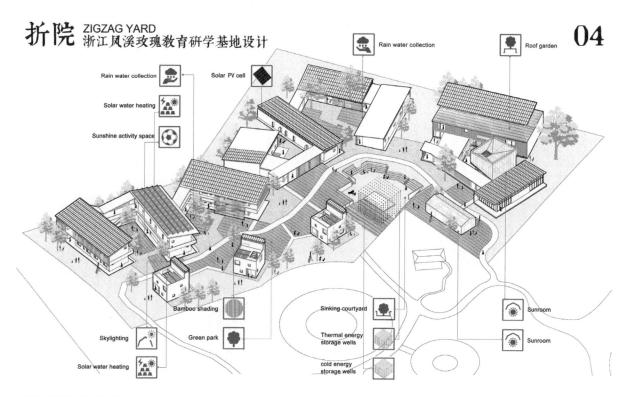

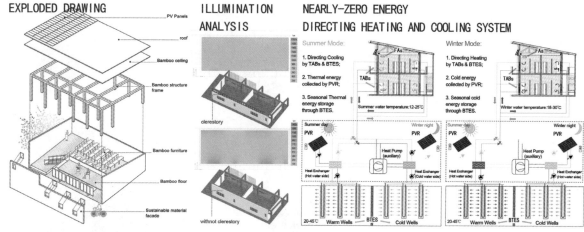

EXPLODED DRAWING

ILLUMINATION ANALYSIS

NEARLY-ZERO ENERGY DIRECTING HEATING AND COOLING SYSTEM

教学中心 teaching center

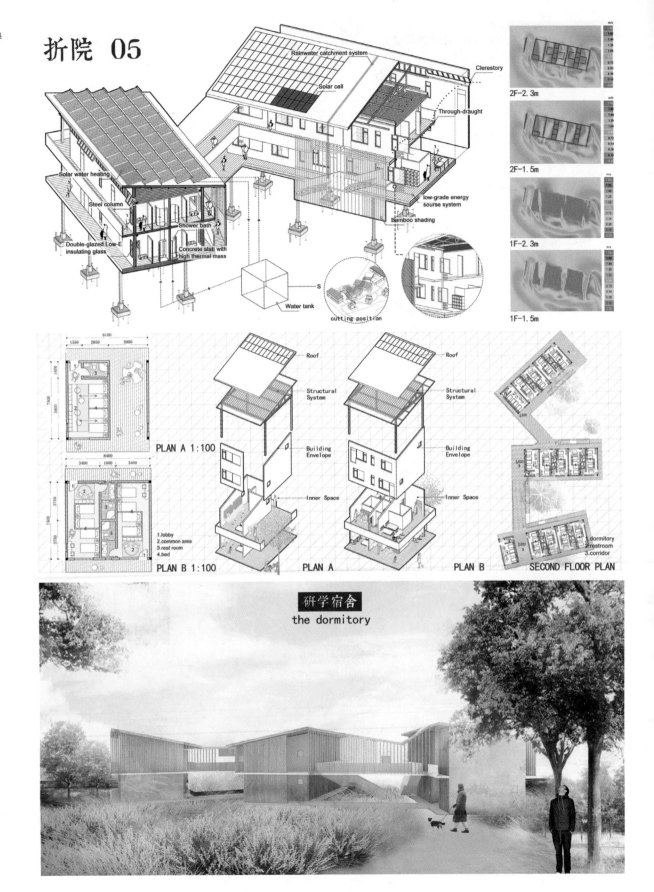

体验单元
experience unit

EXPLODED DRAWING

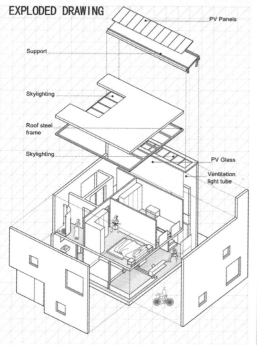

- PV Panels
- Support
- Skylighting
- Roof steel frame
- Skylighting
- PV Glass
- Ventilation light tube

RAIN GARDEN

FIRST FLOOR PLAN 1:100   SECOND FLOOR PLAN 1:100

1. lobby
2. living room
3. dining room
4. kitchen
5. bedroom
6. cloakroom
7. rest room
8. terrace

EXPLODED DRAWING

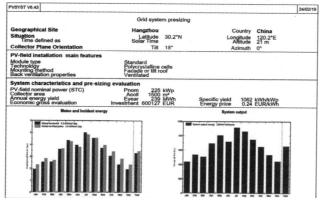

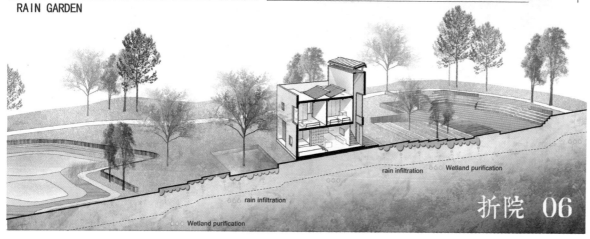

rain infiltration · Wetland purification

折院 06

综合奖·三等奖
General Prize Awarded · Third Prize

注 册 号：6392
项目名称：阳光·风动（凤溪）
　　　　　Sunshine and Wind Power (Fengxi)
作　者：宋汪耀、姚歌、滑维杰
参赛单位：重庆大学
指导老师：周铁军、张海滨

专家点评：

该作品平面分区合理，场地利用充分，采用分散式建筑布局实现了场地内自然通风的有效利用。建筑通过南向的内收式墙面起到一定的遮阳作用，采用种植物屋面、相变蓄热、拔风烟囱等措施，提高室内舒适度。作品在主动式太阳能与建筑结合方面有待提高。

The work plane is reasonably distributed and the site is fully utilized. The decentralized building layout is adopted to effectively use the natural ventilation. The south-oriented inward-retracting wall surface of the building can block the sun to a certain extent, and such measures as a plant roofing, phase change thermal storage and chimney ventilation are employed to improve the indoor comfort level. However, in the combination of active solar energy and buildings, the work still needs to be improved.

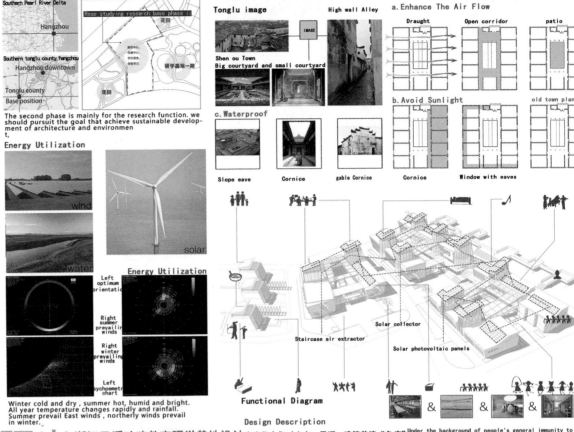

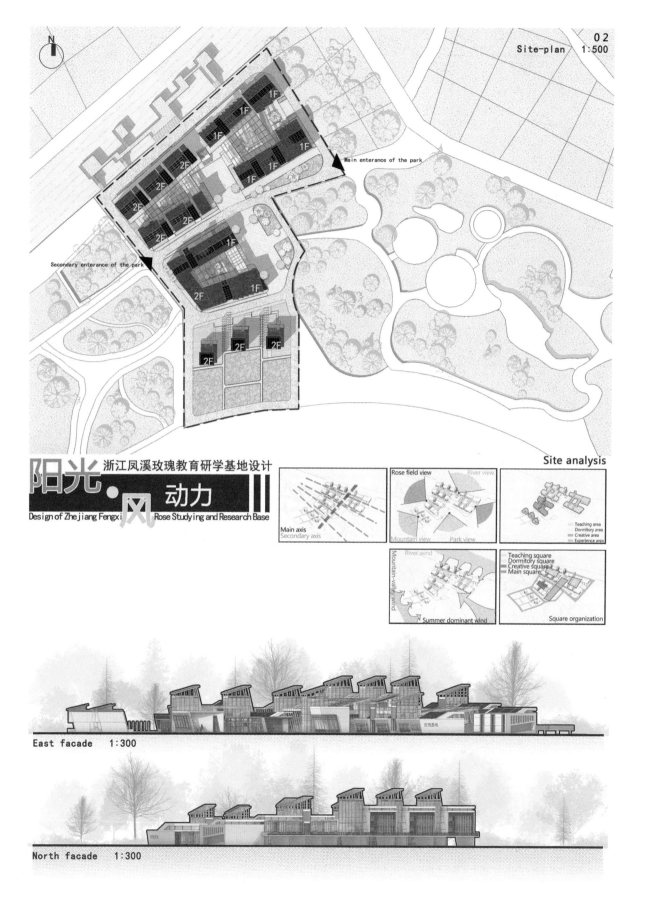

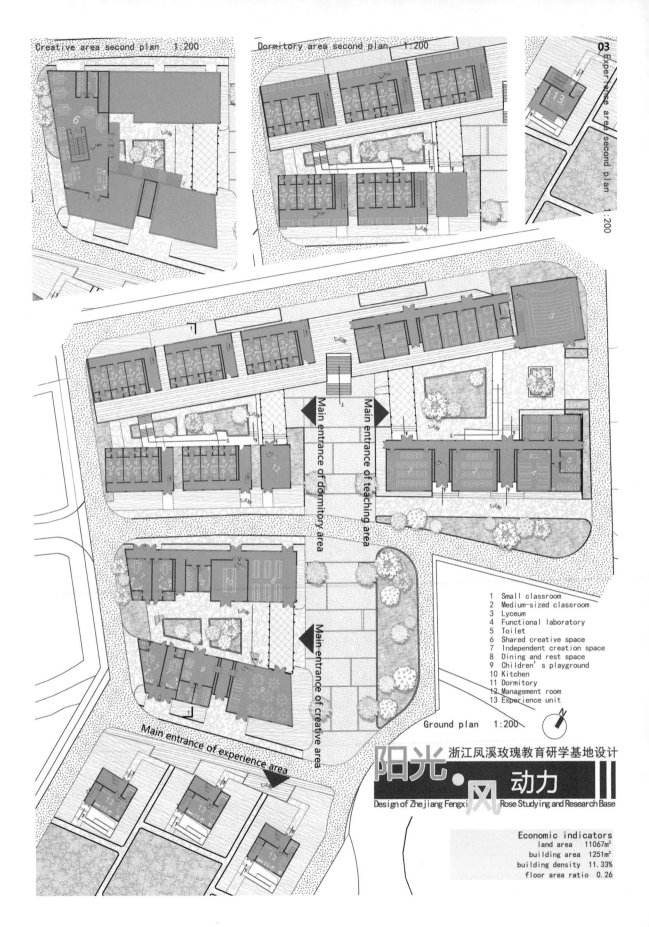

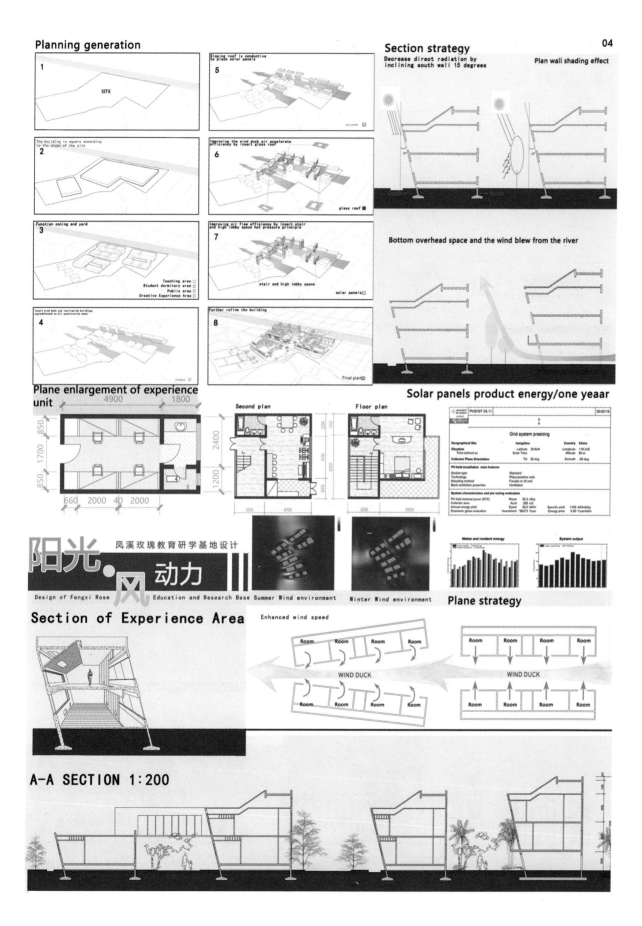

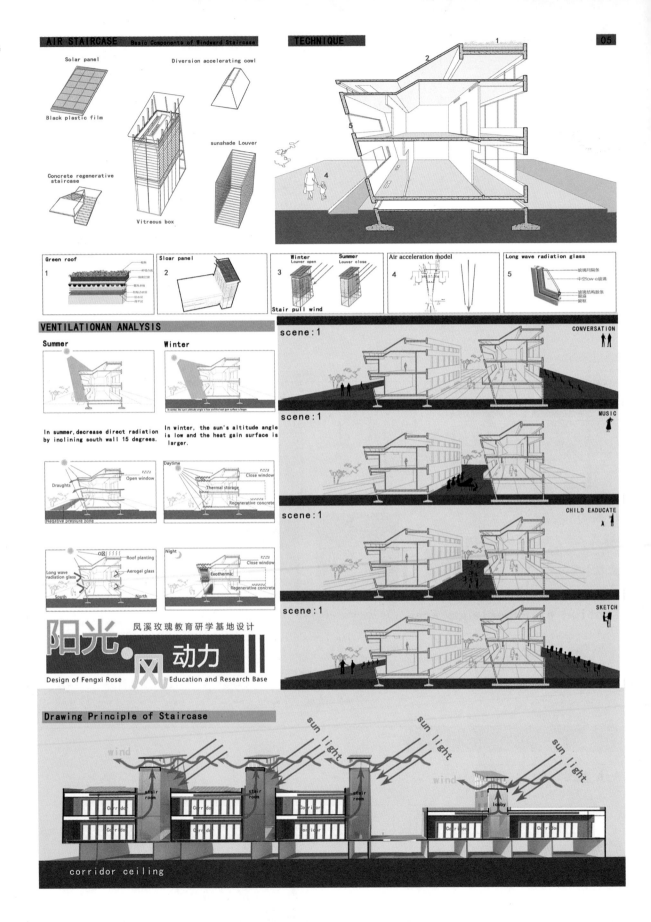

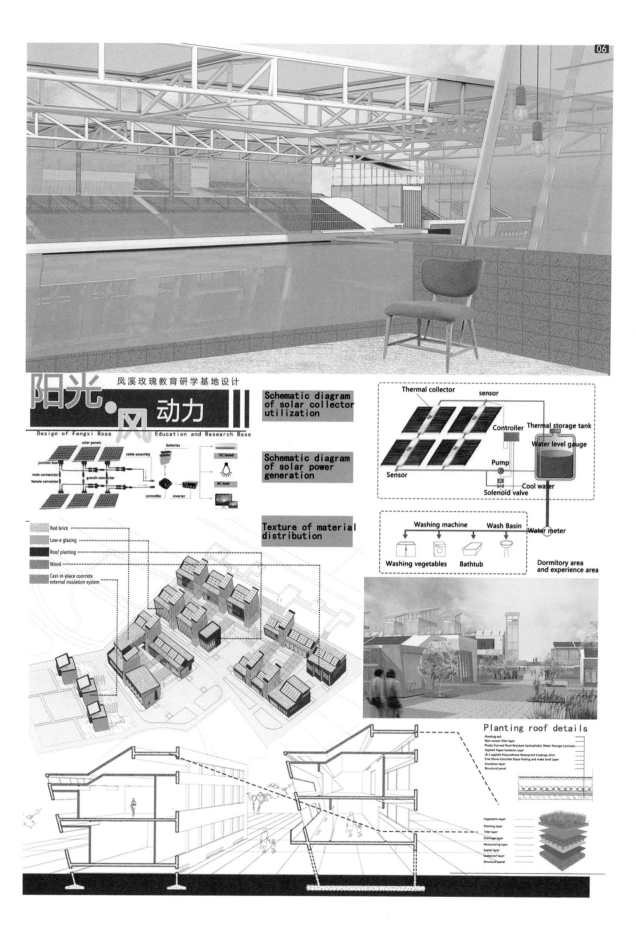

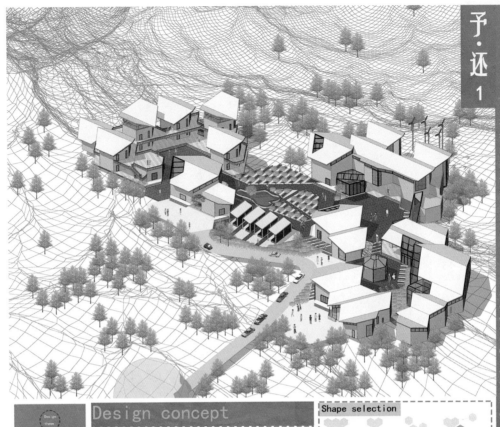

综合奖·三等奖
General Prize Awarded · Third Prize

注　册　号：6612
项目名称：予·还（凤溪）
　　　　　　Cycle（Fengxi）
作　　者：陈梦雪、龚玲莉、黄婉祎
参赛单位：南京工业大学
指导老师：邵继中

专家点评：

该作品以正六边形为基本元素进行布局，围合成三个功能组团，通过主次广场和道路进行连接，功能布局合理，形式有特色。作品采用双层通风墙体，促进室内的通风和采暖，采用地源热泵、沼气、光伏系统等技术，表达深度有待增加。作品中设计的阳光房不宜在该地区使用。

The work takes regular hexagons as basic elements for the layout, and is composed of three functional groups which are connected through main and secondary squares and roads. The function layout is reasonable and the form is distinctive. The work adopts double-layer ventilated walls to improve indoor ventilation and heating, and also uses the ground source heat pump, biogas, photovoltaic system and other technologies, but the expression level still needs to be enhanced. Besides, the sun room designed in the work are inappropriate to use in the local area.

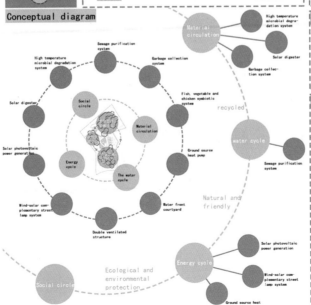

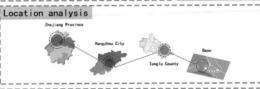

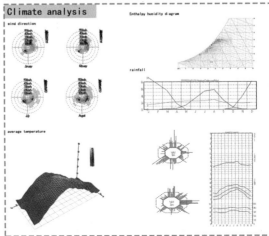

予·还 2

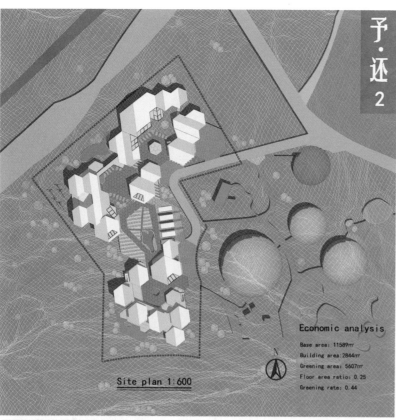

Site plan 1:600

Economic analysis
Base area: 11589㎡
Building area: 2844㎡
Greening area: 5607㎡
Floor area ratio: 0.25
Greening rate: 0.44

Research

Research on traditional houses — Ancient town of Qiangang

→ New rural dwellings
→ Township enterprises
→ Qigang Ancient Town
→ farmland
→ Ligang Reservoir, Mountain

Spatial layout: built according to the water potential, patchwork, do not pursue symmetry.
Architectural form: The main courtyard, residential layout more through the hall, patio.
Construction materials: mainly wood, blue bricks, blue pebbles, LOESS and so on.

Learning from tradition

Fold roof

Water system — Water yard

Courtyard
Design of four main courtyards based on traditional courtyard

Overall Analysis

Functional analysis | Flow analysis | Square analysis | Water body analysis | Field analysis

Water yard placement | Model | Place water yard | Good ventilation

Accommodation unit combination | Creative unit combination | Teaching unit combination

Dormitory monomer analysis

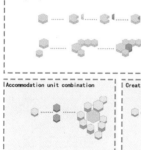

Plan

Reasonable space treatment:
Because the building adopts a hexagonal unit structure, in order to create a more regular space experience for accommodation, equipment is set up to reduce the number of heterosexual rooms in the corner of the building.

Entrance hall analysis

In the entrance hall, the abandoned window frame is processed into a complete display wall, echoing the theme of the architectural design recycling pair.

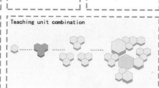

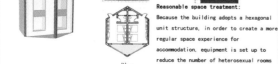 machining — lap

Thermotechnical analysis

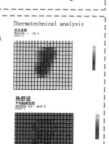

热舒适
平均辐射温度

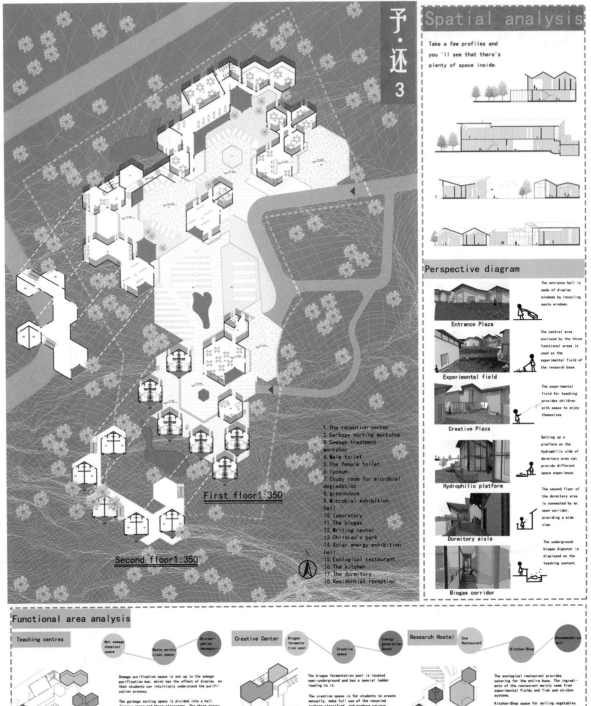
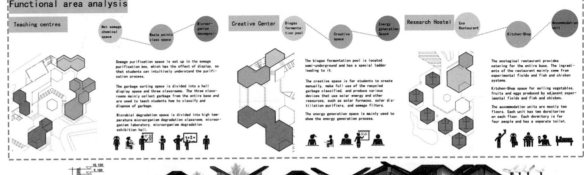

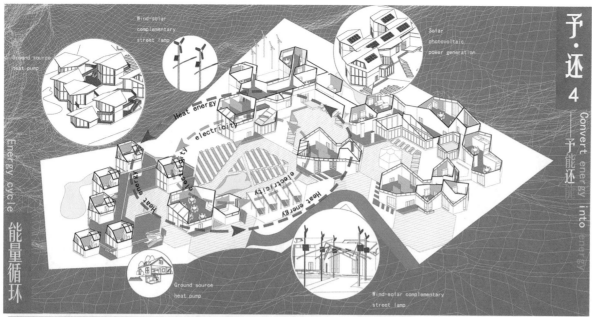
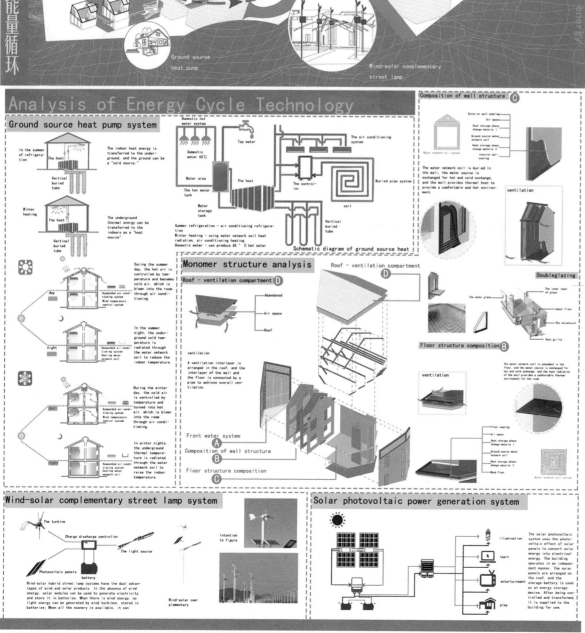

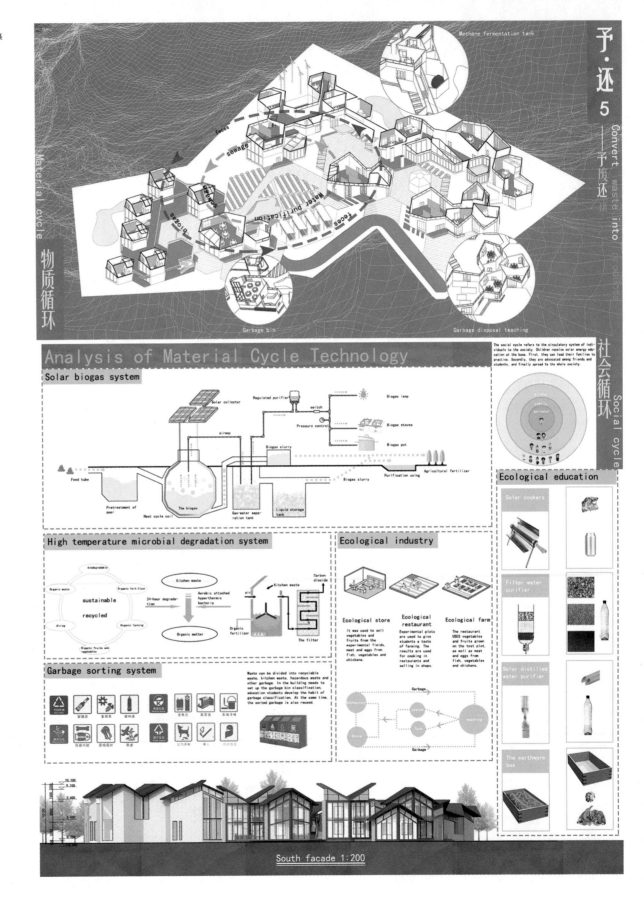

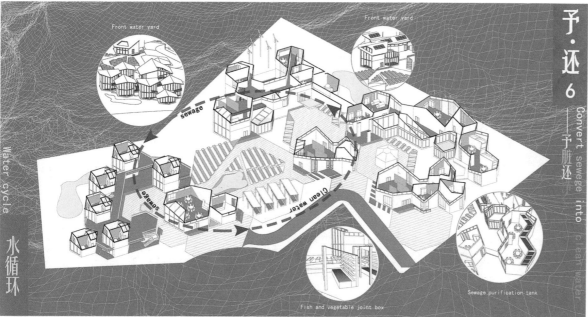

予·还 6 — Convert sewage into clean water

水循环 / Water cycle

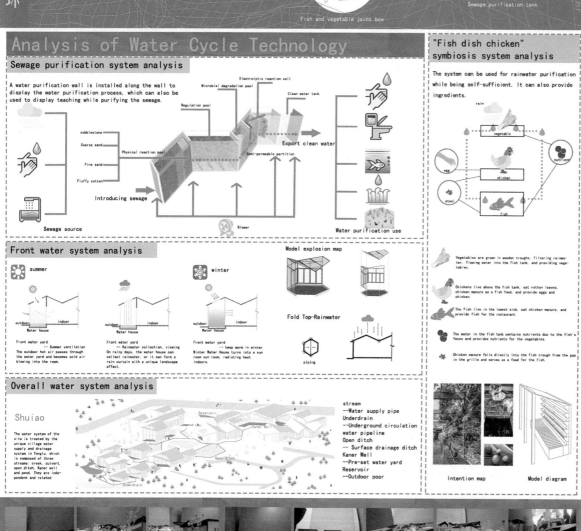

Analysis of Water Cycle Technology

Sewage purification system analysis

A water purification wall is installed along the wall to display the water purification process, which can also be used to display teaching while purifying the sewage.

Front water system analysis

summer
- Front water yard — Summer ventilation
  The outdoor hot air passes through the water yard and becomes cold air blowing into the room.
- Front water yard — Rainwater collection, viewing
  On rainy days, the water house can collect rainwater, or it can form a rain curtain with a unique landscape effect.

winter
- Front water yard — keep warm in winter
  Winter Water House turns into a sun room sun room, radiating heat indoors.

Model explosion map

Fold Top-Rainwater

Overall water system analysis

Shuiao

The water system of the site is treated by the unique village water supply and drainage system in Tonglu, which is composed of three streams: creek, culvert, open ditch, Kaner well and pond. They are independent and related.

- stream
- Water supply pipe
- Underdrain
- Underground circulation water pipeline
- Open ditch
- Surface drainage ditch
- Kaner Well
- Pre-set water yard
- Reservoir
- Outdoor poor

"Fish dish chicken" symbiosis system analysis

The system can be used for rainwater purification while being self-sufficient. It can also provide ingredients.

- Vegetables are grown in wooden troughs, filtering rainwater, flowing water into the fish tank, and providing vegetables.
- Chickens live above the fish tank, eat rotten leaves, chicken manure as a fish food, and provide eggs and chicken.
- The fish live in the lowest sink, eat chicken manure, and provide fish for the restaurant.
- The water in the fish tank contains nutrients due to the fish's feces and provides nutrients for the vegetables.
- Chicken manure falls directly into the fish trough from the gap in the grille and serves as a food for the fish.

Intention map | Model diagram

综合奖·三等奖
General Prize Awarded · Third Prize

注 册 号：6090
项目名称：伫立星河（兴隆）
Standing on the Star River (Xinglong)
作　　者：陈璐、高标、方心怡
参赛单位：南京工业大学
指导老师：林杰文

专家点评：

该作品结合地形布置于山谷底部，建筑朝向有利于采光。活动广场与住宿用房联系便捷。建筑山墙一侧的连续水面和建筑之间的庭院结合，形成了地形高差，与建筑边界组成了形态丰富的半开放活动和交往空间。建筑内部空间合理，设置星空观测平台，突出了酒店主题。生土材料为主的围护结构提高了房屋本体的热惰性，提升版阳光间在北方地区较为可行，对建筑单体的气流组织有较深入的考虑。雨水收集、太阳能热水、光伏和热泵的应用合理。对于山谷区域的雨水汇集和排放还应进一步考虑。

The work is placed at the bottom of the valley in combination with the terrain, and the building orientation is suitable for daylighting. And the event site is easily accessible to the accommodation. The continuous water surface on one side of the gable and the courtyard within the buildings form a topographical height difference, which forms a semi-open space for activities and communication featuring rich forms in combination with the building boundaries. The interior of the building is reasonable in setting, in which a starry sky observation platform is set to highlight the hotel theme. The

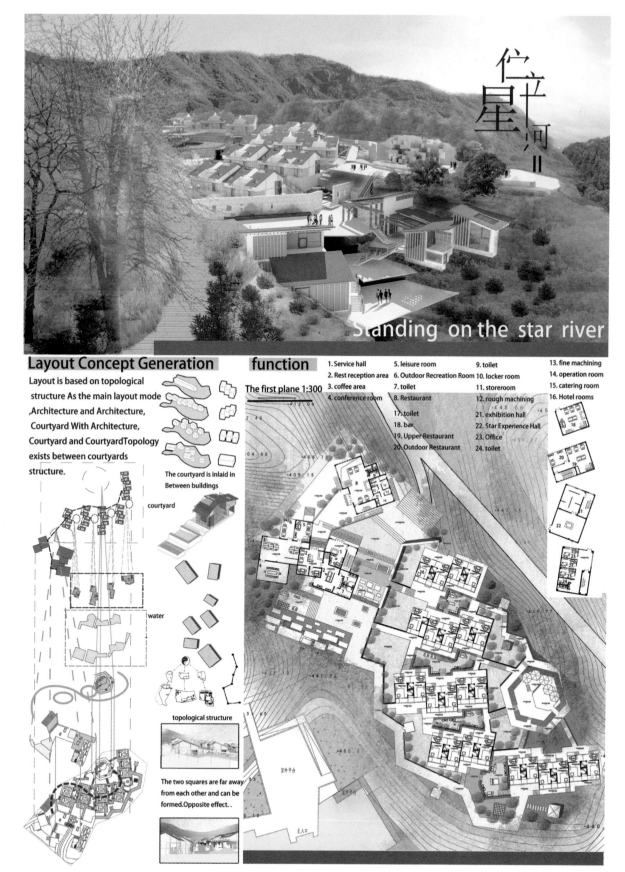

building envelope mainly decorated with earth-based material improves the thermal inertia of the building body, so it is feasible to upgrade the sunlight level within rooms in the north of China. In addition, more deeper considerations are given for the airflow organization of the building unit. Besides, the application of rainwater harvesting, solar hot water, photovoltaics and heat pumps is reasonable. However, rainwater collection and discharge in the valley area should be further considered.

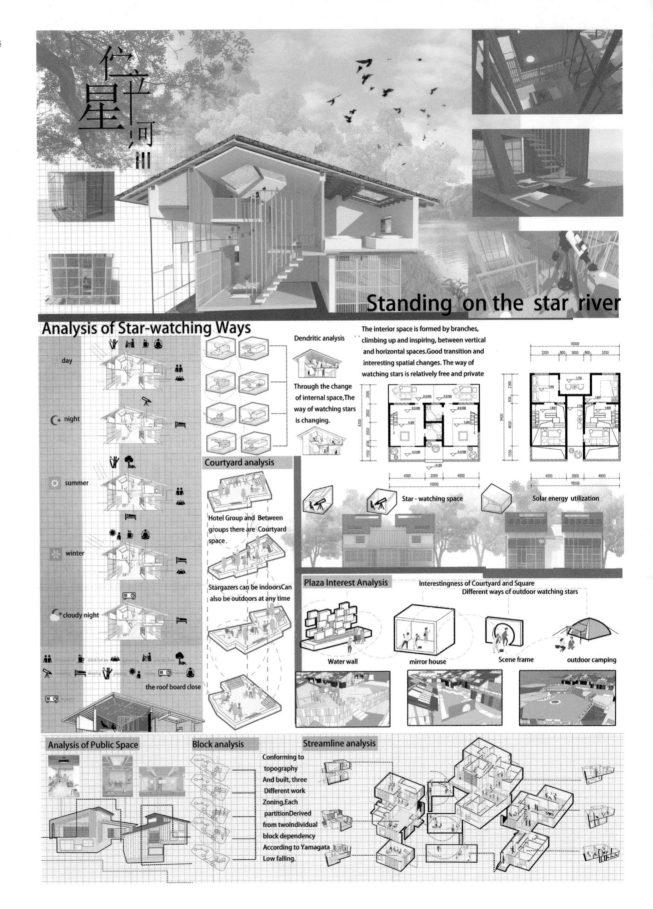

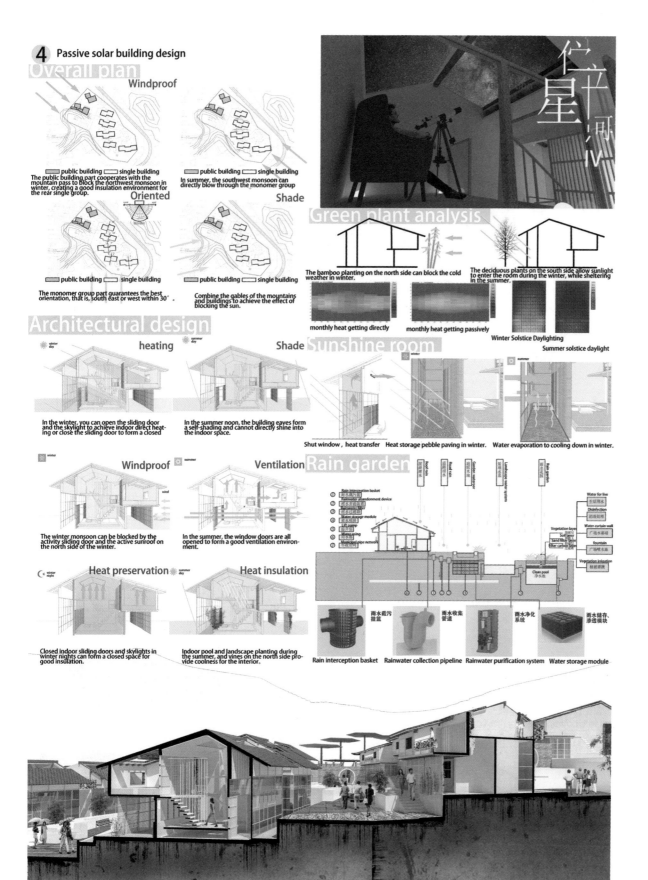

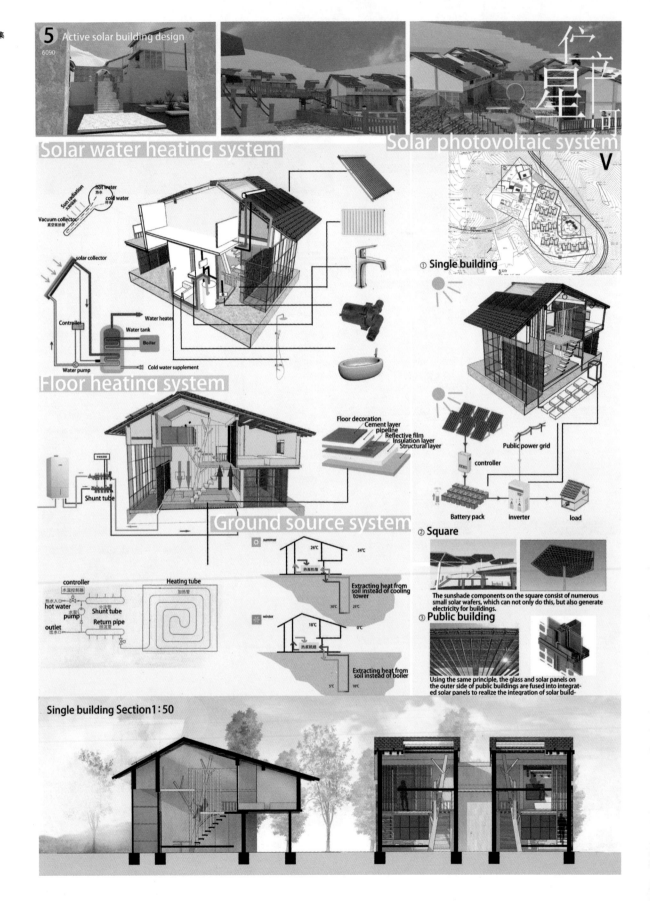

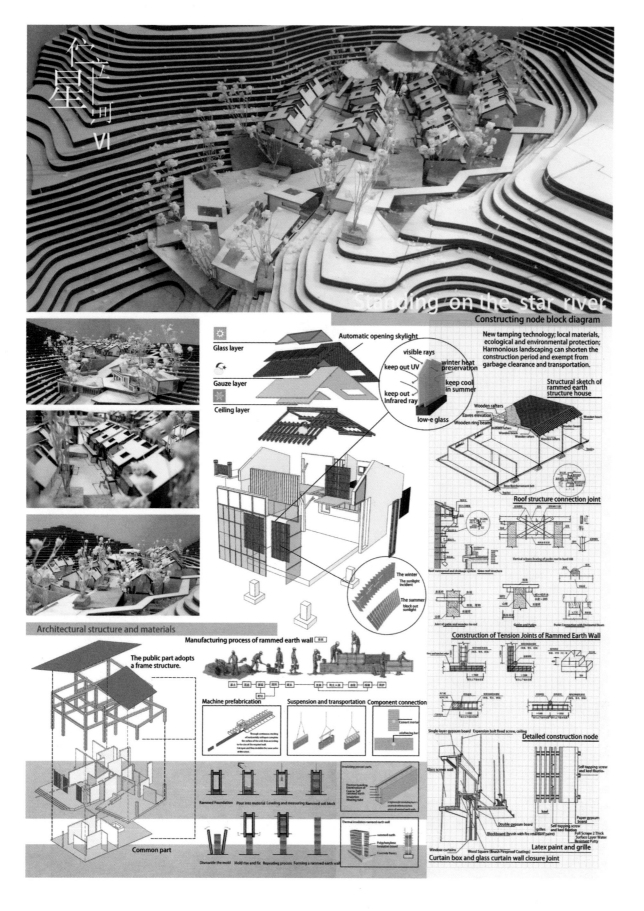

2019台达杯国际太阳能建筑设计竞赛获奖作品集

RAISING IN THE WALL 1
墙·袱

综合奖·三等奖
General Prize Awarded · Third Prize

注 册 号：6156
项目名称：墙袱（兴隆）
　　　　　Raising in the Wall (Xinglong)
作　　者：许锦灿、史小宇、吴佳晋、周浩
参赛单位：南京工业大学
指导老师：张伟郁、杨亦陵

专家点评：

该作品建筑紧邻红线，结合地形成组环绕布置，让出谷底部分形成围合式布局。充分利用场地有利于该地区汇集雨水的排放。多个住宿单元为一组，分五组布置，形成小规模居住单元的同时避免孤独感，适合科考、会议接待和旅游度假的使用模式。建筑单体设计充分考虑了利用多种方式对室内气流的合理组织。太阳能空气集热结合卵石蓄热、太阳能烟囱加强通风等被动技术和光伏发电、太阳能热水等主动技术也得到较好应用。

The building is adjacent to the red line and forms a surrounding group in combination with the terrain, and the bottom of the valley is overhead to form a closed layout. To make full use of the site is conducive to emitting the rainwater collected in the area. Several accommodation units are arranged in a group, with a total of five groups set, forming a small-scale living unit while avoiding loneliness, so that is a proper model for scientific research, conference reception and travel. The single building design fully considers the rational organization of indoor airflow in various ways. Moreover, passive technologies such as solar air heat collection combined with

DESIGN NOTES

由于场地夏季凉爽湿润，冬季寒冷干燥。因此我们在采取组团式布局的同时，通过挡风墙的围合以抵挡严寒。

取"墙"喻之为被，将客房单体一组，紧紧扎在一起，使人如同襁褓中的婴儿一般，倍感温暖。在客人从外到内进入到客房的过程中，"广场—廊院—平台"的观星序列依次展开，私密性逐渐加强，空间体验丰富。

在节能设计上，则利用山地优势，借鉴当地传统民居采暖做法，节能而高效。

The local climate is cool in summer and cold and dry in winter. Accordingly, we took the form of group layout, and surrounded by the windshield wall, to withstand the cold, take the "wall" metaphor as "swaddling clothes", the room unit-blanket tied together tightly, like a baby in swaddle, feel warm a lot. As guests enter the guest room from outside to inside, the stargazing sequence of the square-courtyard-platform expands sequentially, and the privacy is gradually enhanced and the experience is hierarchical.
In terms of energy-saving design, the advantages of mountainous areas and local traditional practices are utilized, to achieve the aim of energy-saving and high-efficiency.

SITE LOCATION INFORMATION

Chengde City

SITE METEOROLOGICAL

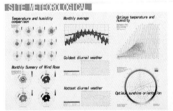

LOGIC OF SITE LAYOUT

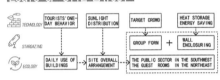

SITE SUNSHINE

The Summer Solstice solar radiation range
The Winter Solstice solar radiation range

CROWD CONFIGURATION

ALONE / COUPLE / CLUB / FAMILY

Users have the characteristics of diversity and collectivity, and therefore, the use of diversified group layouts is considered in the subsequent scheme generation.

ONE-DAY BEHAVIOR

Activity / Food / Stargazing / Rest
0h — 24h

BUILDING DURATION
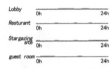
Lobby / Restaurant / Stargazing site / guest room
0h — 24h

The public part starts to use more in the morning, the guest room begins to use more in the evening.
The former should face east, the room should face west, in order to win more sunshine.

GENERAL PLAN

SITE MAIN ENTRANCE
LOBBY ENTRANCE
RESTAURANT ENTRANCE
LOGISTICS ENTRANCE
SERVICE HALL
RESTAURANT
STARGAZING ROOM

GROUP GENERATION MIND MAP

 CAMPSITE
Enhance the communication
Gather the observers

 HUDDLE
Be conducive to the formation suitable climate.

 QUADRAN
Be extracted to enclose multiple units.

 NORTH-SLOPE
Wind barriers are used to create microclimate

 **FOLDED**
Adjust the angle of wind barriers by software analysis

 **OPENING**
Open the south side to improve ventilation

 **ADJUSTMENT**
Adjust the height of the wind barriers

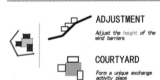 **COURTYARD**
Form a unique exchange activity place

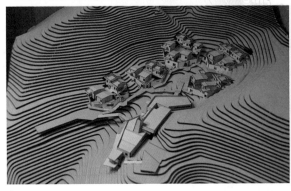

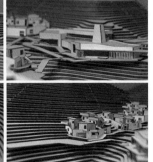

RAISING IN THE WALL 2
墙·裸

pebble heat storage, ventilation-enhancing solar chimney, as well as the active technologies including photovoltaic power generation and solar hot water are also well applied.

SITE ANALYSIS
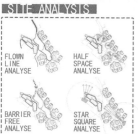
- FLOWN LINE ANALYSE
- HALF SPACE ANALYSE
- BARRIER FREE ANALYSE
- STAR SQUARE ANALYSE

STARGAZE ORGANIZATION

COMMUNITY — A Star-watching Square is set at each end of the site.

GROUP — Constructions enclosed to form a star-watching site.

MONOMER — Interpenetration of Star-watching Site and Interior Space.

WALL LECTOTYPE

Changing the shape and angle of the wind retaining wall can optimize the wind direction and low wind pressure area where the winter wind blows through the wall.

SITE PLAN 1:500

Economic Technological Index
- Building area : 2545㎡
- Greening rate : 74.2%
- Covered area : 1810㎡
- Construction land rate : 7927㎡
- Floor area ratio : 0.32
- Single room : 5
- Double room : 12
- Double suite : 7

LOCATION ANALYSIS

WALL GRNERATING

The winter wind direction of the site is northwest wind. And the east and west valleys form a tunnel effect, increase wind speed and accelerate heat loss in the site.

The summer wind direction of the site is southeast wind, but it is blocked by the valley on the southeast side, forming a vortex effect, which makes the ventilation in the site poor.

 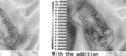

After adding the building group, it is intended to form a microclimate conducive to heat preservation between the building clusters. But due to the building interval, the wind speed is slow.

With the addition of building clusters and wind retaining walls, the summer monsoon will be further weakened, so that ventilation within the group cannot be carried out through the monsoon.

The wind retaining wall outside the group has remarkable blocking effect on winter monsoon. The decrease of wind speed forming a local microclimate with heat preservation effect.

Properly cancel the wind retaining wall on the southwest side of the group, or set up ventilation vents to guide the summer monsoon, and strengthen the ventilation effect.

The angle and height of wind retaining wall can be adjusted by combining lighting and visual field, so as to achieve better effect of adjusting microclimate.

Through unidirectional ventilation louver and green setting, the ventilation opening of retaining wall can be reduced to guide the winter wind.

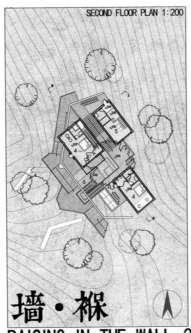
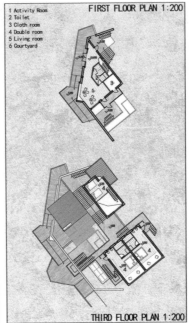
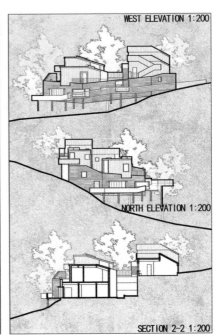

墙·裸
RAISING IN THE WALL 3

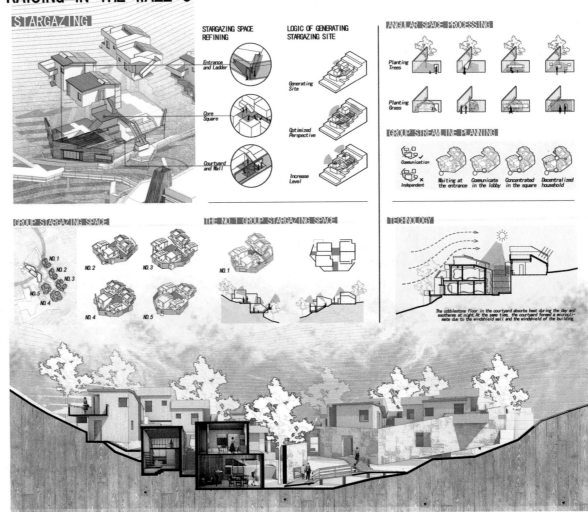

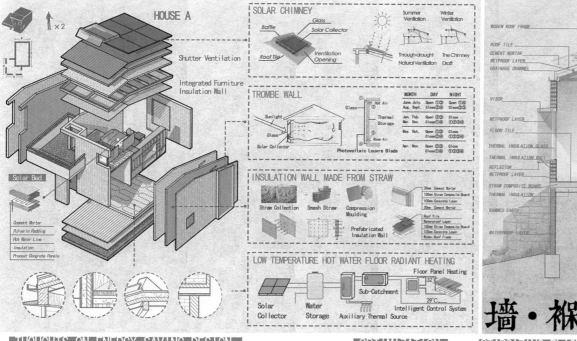

RAISING IN THE WALL 4

墙·褓

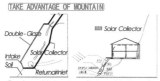

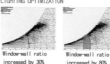
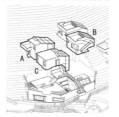

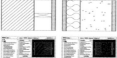

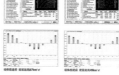

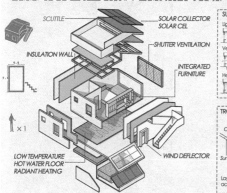

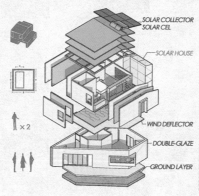
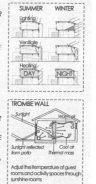

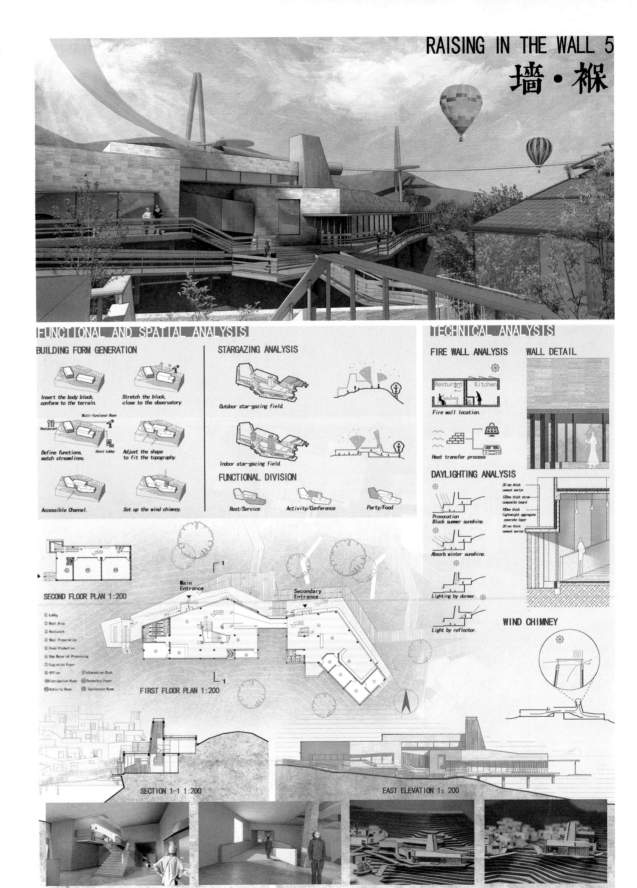

RAISING IN THE WALL 6
墙·槑

ECOLOGICAL EFFECTS

   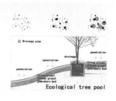

1. Wall storing water
2. Wall protecting saplings
3. Micro-climate formation

Ecological tree pool

ACTIVE SOLAR TECHNOLOGY

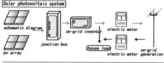

Solar photovoltaic system

Photovoltaic power converts light energy directly into electrical energy. The roof of the building is placed with a suitable angle of photovoltaic panels. The collected solar energy is used for building electricity, self-sufficient, and the equipment is quiet and has a long life. Recycled rainwater are used for cleaning and cooling of photovoltaic systems, in order to improve the system

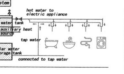

Solar water heater system

Domestic hot water is an indispensable part of life. Flat-panel solar water heaters are simple to install, have a small footprint, aren't limited by the installation environment, and can be installed on balconies, windows, roofs, etc. And it has a high heat extraction rate, is very energy efficient, and is easy to use.

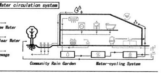

Water circulation system

The campsite adopts the Low Impact Development Concept (LID) to save every drop of water in the house. The system consists of the Water-cycling System and the Community Rain Garden. The roof rainwater collection mixed with grey water for filtration and reverse osmosis treatment to alleviate water shortages and flood waters drainage pressure. And Each unit is linked to the community's water recycling system.

Environment control system

The campsite uses small integrated high efficiency HVAC system to avoid the many performance issues. The comfort system not only ensures pleasant interior environment but also reduces the energy consumption of the house. The intelligent housekeeper masters a machine learning algorithm (AI), which calculates the most energy-efficient equipment operation mode and regulates it through real-time data and user feedback.

PREFABRICATION DESIGN

1. Transportation
2. Assembly Splicing
3. Principal structure generation
4. Wall structure and roof
5. Prefabricated panels and roof insulation
6. Facilities and veneers

综合奖·三等奖
General Prize Awarded · Third Prize

注 册 号：6314
项目名称：卧看星河尽意明（兴隆）
　　　　　Lying in The Galaxy
　　　　　（Xinglong）
作　　者：夏毓翎、赵　雨、朱　曦、
　　　　　鲍慧敏
参赛单位：南京工业大学
指导老师：罗　靖、杨亦陵

专家点评：
该作品采用组团式布局，根据山地高差分布在场地内，与场地结合较好，场地内空间变化多样，交通流线较为合理。建筑单体造型丰富，被动式太阳能采暖、采光等技术应用合理。对于主动式太阳能技术及其他技术的设计有所欠缺。

The work adopts a group layout, which is distributed in the site according to the mountain height difference, and is well integrated with the site. The indoor space of the site varies and the traffic flow is rather reasonable. The single building boasts abundant shape, and such technologies as passive solar heating and lighting are used reasonably. Nevertheless, the design capacity for active solar technology and other technologies is rather insufficient.

卧看星河尽意明 01
Lying in the galaxy

卧看星河尽意明 02
Lying in the galaxy

- **SITE SYNTHESIS ANALYSIS**

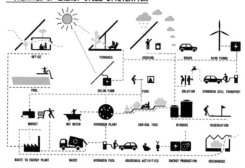

- **OVERVIEW OF ENERGY CYCLE UTILIZATION**

- **ARTIFICIAL LIGHT ENVIRONMENT TREATMENT**

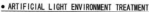

- **SITE WIND ENVIRONMENT ANALYSIS**

Wind speed analysis

Wind pressure analysis

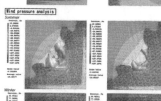

- **NATURAL WATER SEEPAGE SYSTEM**

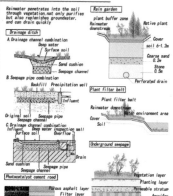

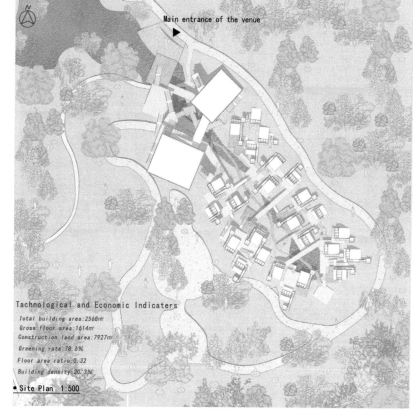

Main entrance of the venue

Technological and Economic Indicaters
Total building area: 2568m
Gross floor area: 1614m
Construction land area: 7927m
Greening rate: 78.6%
Floor area ratio: 0.32
Building density: 20.3%

- **Site Plan 1:500**

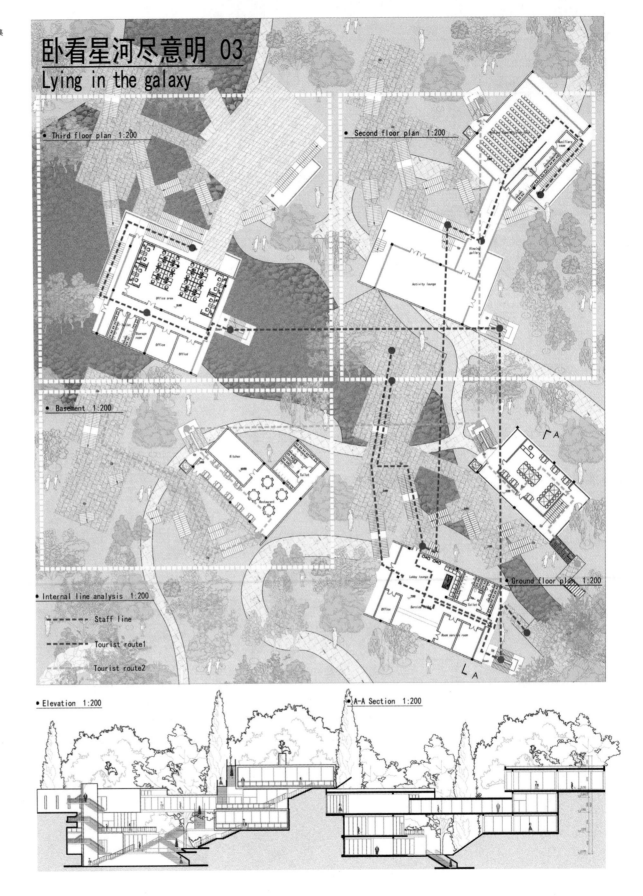

卧看星河尽意明 04
Lying in the galaxy

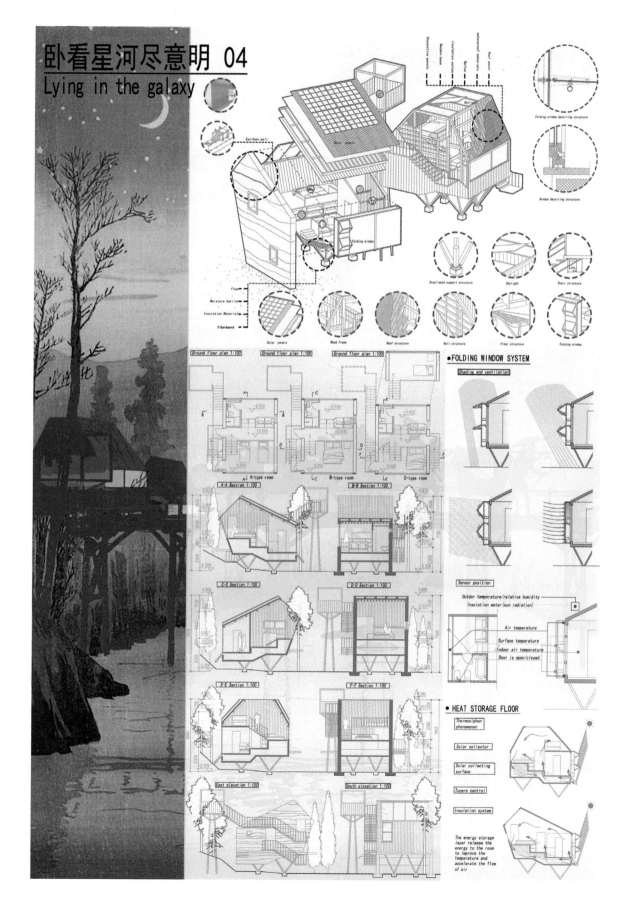

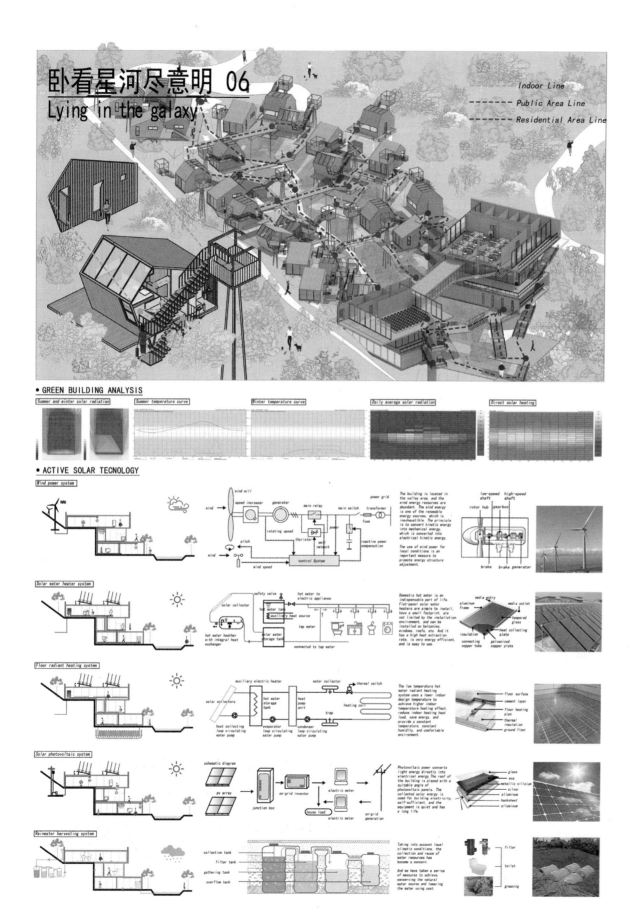

综合奖 · 优秀奖
General Prize Awarded · Honorable Mention Prize

注 册 号：5745
项目名称：一方（凤溪）
　　　　　Nostalgia（Fengxi）
作　　者：寇朦心、康夏田、王幸珂、
　　　　　马　艳、张　茜
参赛单位：西安科技大学
指导老师：孙倩倩

构建围合与穿透空间的对话关系　一方
Construct the dialogue relationship of enclosure and penetrative space

SITE LOCATION ANALYSIS

Hangzhou　　　　TongLu Powiat

The project site is located in Fengxi Rose Education and Research Base, Sanxincun, south of Fengchuan Street, Tonglu County, Hangzhou City, Zhejiang Province. The project site has dense vegetation and beautiful surface landscape.

■ DESIGN DESCRIPTION

一方水土养一方人，我们希望寻求一种人文、生态可持续的乡土生存状态唤起青少年的记忆，重视家乡文化传承，关注自己内心根源，不管他们来自何处，身处何方。
设计采用传统天井和庭院式布局，建筑通过院落组织。木格栅遮阳连廊，强化功能联系，同时界定室内外空间体验序列。被动式阳光房、空心砖墙、遮阳板、热压通风等，用于解决冬季保温及夏季防热问题；屋面太阳能集热系统、雨水收集、地板热辐射采暖等主动式技术实现低投入高能效节能。

A place where you live makes a man who you are. We hope to seek a humanistic and ecologically sustainable local living condition to arouse the memory of teenagers, which attach importance to the cultural heritage of their hometown, and pay attention to inner roots. No matter where they come from or where they are. This design adopts the traditional patio and courtyard layout, each building organized through the courtyard and connects by wooden grid sunshade corridor, which not only strengthens the function of connection, but also defines the experience sequence of the exterior space. Passive design we use sunshine house, hollow brick, sun visor, hot-press ventilation to solve the problem of winter insulation and summer ventilation. Also through solar roof collector system, rainwater collection, radiant floor heating etc active designs achieve a low investment, high efficientcy way.

Features Of Regional Architecture

Courtyard　　　　　　　　　Slope Roof

Corridor　　　　　　　　　　Grey Space

HangZhou Climate Analysis

Winds Analysis　　Prevailing Winds　　Optimum Orientation　　Maximum Wind Temperature

Diffuse Summary　　Direc Solar Radiation　　Max temperature & Wind(summer)　　Max temperature & Wind(winter)

02

构建围合与穿透空间的对话关系 一方
Construct the dialogue relationship of enclosure and penetrative space

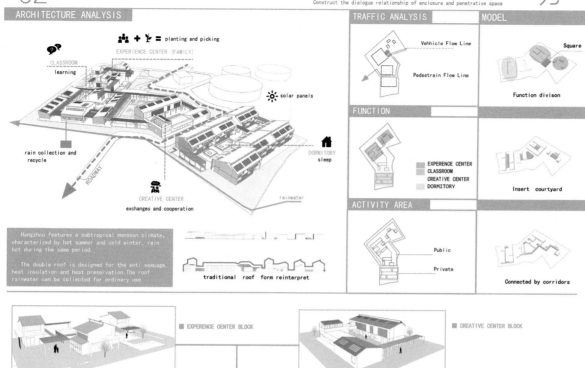

Ventilate Analysis

Decomposition Of Energy-Saving

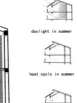

Sunlight Simulation

Shadow Analysis

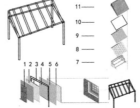

CLASSROOM STRUCTURE

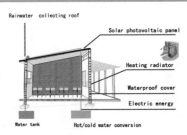

WEST ELEVATION 1:300

NOSTALGIA

1. EXPERENCE CENTER
2. SMALL CLASSROOM
3. MIDDLE CLASSROOM
4. LYCEUM
5. BATHROOM
6. RESTAURANT
7. CHILDREN'S ACTIVITY ROOM
8. SMALL ACTIVITY ROOM
9. GALLEY
10. DISINFECTION ROOM
11. KITCHEN
12. BATHROOM
13. LARGE ACTIVITY ROOM
14. BATHROOM
15. BEDROOM

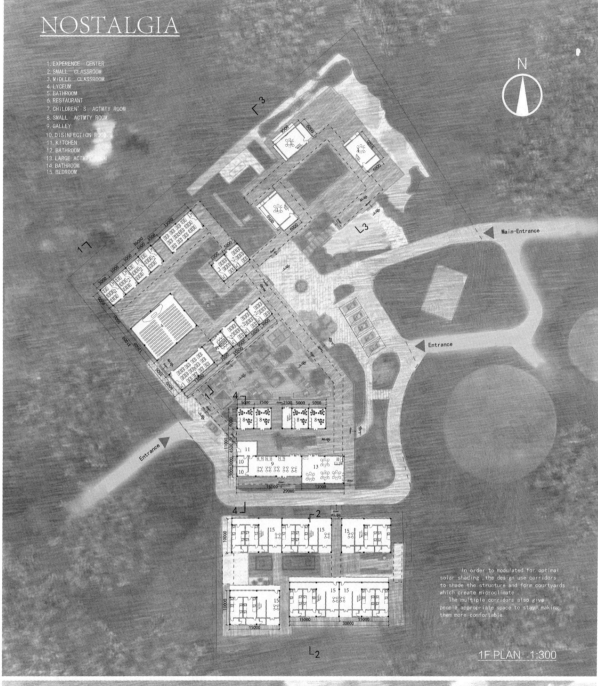

In order to modulated for optimal solar shading, the design use corridors to shade the structure and form courtyards which create microclimate.
The multiple corridors also give people appropriate space to stay, making them more comfortable.

1F PLAN 1:300

EAST ELEVATION 1:300

05 构建围合与穿透空间的对话关系 一方
construct the dialogue relationship of enclosure and Penetrative space

■ 宿舍单元放大
Domitory Unit

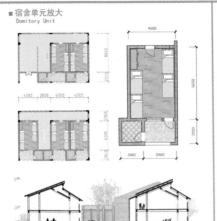

DOMITORY SECTION 2-2 1:200

CREATIVE CENTER SECTION 4-4 1:200

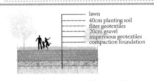
Rain gargen detail of construction 1:30

EXPERIENCE CENTER SECTION 3-3 1:200

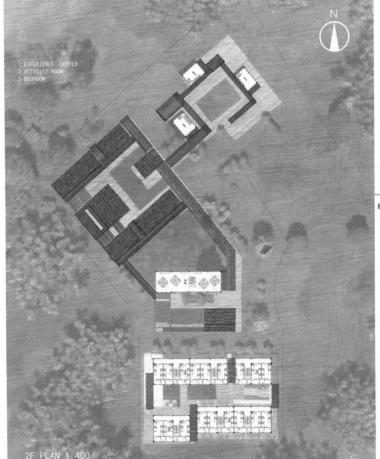

1. EXPERIENCE CENTER
2. ACTIVITY ROOM
3. BEDROOM

2F PLAN 1:400

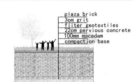
floor decoration detail of construction

plaza brick
3cm grit
filter geotextiles
22cm pervious concrete
100mm macadam
compaction base

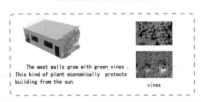
The west walls grow with green vines. This kind of plant economically protects building from the sun.   vines

■ Planting modular wall

Metal frame
Polyamide Fiber Layer 30mm (Plant growth layer)
Polyvinyl chloride layer 20mm (Root barrier layer)

Composition of Unit Module
Bolted connection
Connection between modules

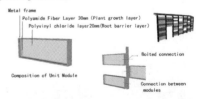

经济技术指标 Economic Technological Index			
占地面积 covered area	11700m²	总建筑面积 overall floorage	2870m²
停车场 Parking		建筑面积 Building area	1631m²
容积率 floor area ratio	0.22	绿地率 Greening rate	53%
建筑密度 building density	15%	建筑层数 building altitude	1F、2F
体验区 Experience area	450m²	宿舍区 The dormitory area	1200m²
体验馆 Experience area	500m²	餐厅面积 The restaurant area	120m²
独立创作空间 Independent space	125m²	儿童乐园面积 Children's area	100m²
共享创作空间 Shared creative space	300m²		

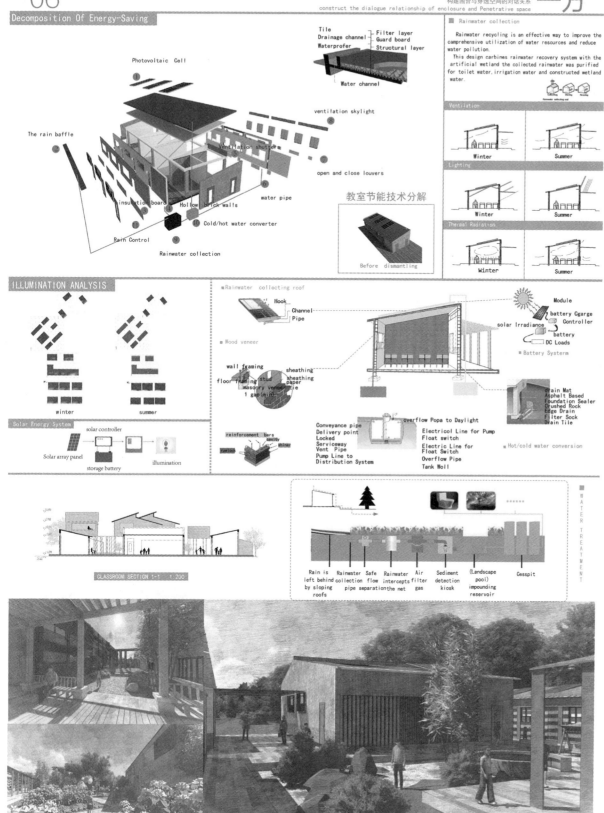

风檐寸暑 FENGXI ROSE EDUCATION RESEARCH BASE PROJECT 01

综合奖·优秀奖
General Prize Awarded·Honorable Mention Prize

注 册 号：5767
项目名称：风檐寸暑（凤溪）
　　　　　Wind Eaves Inch Sundial
　　　　　(Fengxi)
作　　者：杨　娜、李　伟、马昌华、
　　　　　杨佳瑶、黄锐聪、王　兰
参赛单位：西安科技大学
指导老师：孙倩倩

● Geography and climate analysis

Good sunshine: conducive to the use of solar, to meet the needs of the elderly buildings
Perfect temperature: no hotsummer, nocold winner. The difference of temperature between day and night is small
Bad humidity: heavy air humidity, easy to get mold
Nice wind: moderate speed, pleasant breeze

COMFORT LEVEL OF HANG ZHOU

● Ecotect analysis

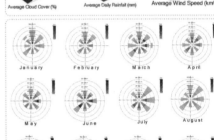

● Historical and cultural analysis

environmental pollution | The horsehead walls | White walls and black tiles | Slope roof

Shadow play | Shaoxing opera | Tea plucking | Changxing baiye dragon

● Hangzhou climate analysis

upper left: best orientation
upper right: diurnal averages
lower left: psychrometric chart
lower right: prevailing winds

Wind Frequency (Hrs) | Average Wind Temperatures | Average Relative Humidity | Average Rainfall (m)

杭州夏季炎热、湿润，冬季寒冷、干燥，春秋两季气候宜人；干湿季节分明，秋末冬出少雨，冬季初春多雨

Hangzhou is hot and humid in summer, cold and dry in winter. Spring and autumn two seasons pleasant climate, dry and wet season clear, late autumn winter out of less rain, winter early spring rainy

January | February | March | April
May | June | July | August
September | October | November | December

● 该项目位于浙江凤溪，方案选用符合当地特色的青瓦白墙元素进行设计，根据功能的不同进行体块的组合细化，通过气候风向等因素进行体块排布。项目整体通过庭院进行空间组织，若干庭院空间的串联实现了建筑与周围环境，以及建筑内部各部分功能空间的有机组织，庭院空间的收放、闭合、引导使建筑群体保持了应有的开放性与私密性，空间趣味性更强。该项目在被动式绿色节能设计的同时采用大量先进的主动式节能技术，将建筑与技术相融合，使教育基地做到真正的绿色节能，为孩子们提供舒适良好的学习成长环境。

● The project is located at Fengxi, Zhejiang Province. The project chooses the elements of black tile and white wall which are in line with local characteristics to design. According to different functions, the block combination is refined, and the block arrangement is carried out through factors such as climate, wind direction and so on. The whole project is organized through the courtyard, and the series of several courtyard spaces realizes the organic organization of the building and the surrounding environment, as well as the functional space of each part of the building. The opening, closing and guiding of the courtyard space makes the building group keep its due openness and privacy, and the space interesting is stronger. The project adopts a large number of advanced active energy-saving technologies while passive green energy-saving design. It integrates building and technology, makes the education base achieve real green energy-saving, and provides a comfortable and good learning and growing environment for children.

风檐寸晷 FENGXI ROSE EDUCATION RESEARCH BASE PROJECT 02

2019 台达杯国际太阳能建筑设计竞赛获奖作品集

● Site-plan general layout 1:500

● Base on analysis

The thrust of the base shape

Original road of the base

Base main wind direction

The best direction

● General analysis

Withdrawal of building red line / The car line

Follow the southeast wind

Optimal daylight orientation

● Block analysis

Streamline arrangement / Spatial Sequence Arrangement

Architectural Form and Shape Determination

Energy-saving Technology Transportation

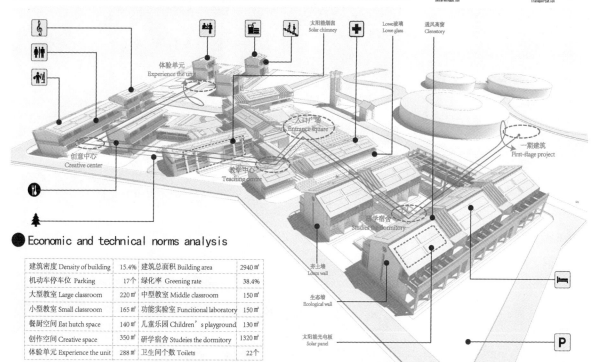

● Economic and technical norms analysis

建筑密度 Density of building	15.4%	建筑总面积 Building area	2940㎡
机动车停车位 Parking	17个	绿化率 Greening rate	38.4%
大型教室 Large classroom	220㎡	中型教室 Middle classroom	150㎡
小型教室 Small classroom	165㎡	功能实验室 Functional laboratory	150㎡
餐厨空间 Eat hutch space	140㎡	儿童乐园 Children's playground	130㎡
创作空间 Creative space	350㎡	研学宿舍 Studeies the dormitory	1320㎡
体验单元 Experience the unit	288㎡	卫生间个数 Toilets	22个

077

风檐寸暑 FENGXI ROSE EDUCATION RESEARCH BASE PROJECT

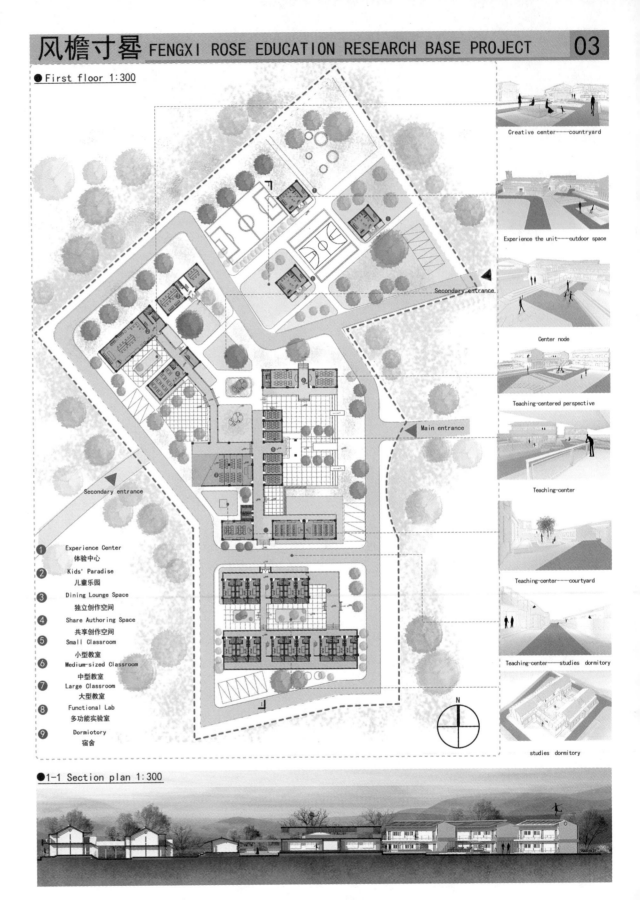

● First floor 1:300

1. Experience Center 体验中心
2. Kids' Paradise 儿童乐园
3. Dining Lounge Space 独立创作空间
4. Share Authoring Space 共享创作空间
5. Small Classroom 小型教室
6. Medium-sized Classroom 中型教室
7. Large Classroom 大型教室
8. Functional Lab 多功能实验室
9. Dormiotory 宿舍

Creative center——countryard

Experience the unit——outdoor space

Center node

Teaching-centered perspective

Teaching-center

Teaching-center——courtyard

Teaching-center——studies dormitory

studies dormitory

● 1-1 Section plan 1:300

风檐寸晷 FENGXI ROSE EDUCATION RESEARCH BASE PROJECT 04

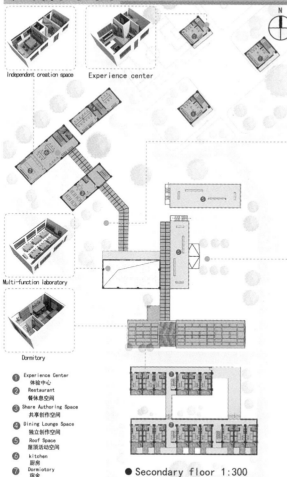

- Independent creation space
- Experience center
- Multi-function laboratory
- Dormitory

1. Experience Center 体验中心
2. Restaurant 餐休息空间
3. Share Authoring Space 共享创作空间
4. Dining Lounge Space 独立创作空间
5. Roof Space 屋顶活动空间
6. kitchen 厨房
7. Dormiotory 宿舍

● Secondary floor 1:300

● Tecnonic node

Staircase Node Detail   Slope Roof Gutter Detail

● East elevation 1:300

● School group perspective

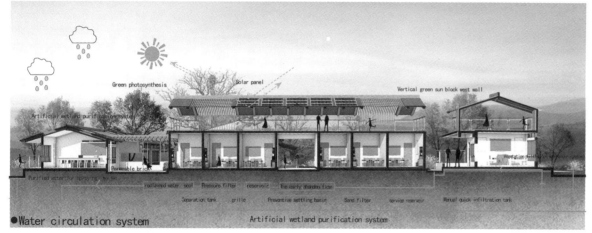

● Water circulation system

风檐寸暑 FENGXI ROSE EDUCATION RESEARCH BASE PROJECT

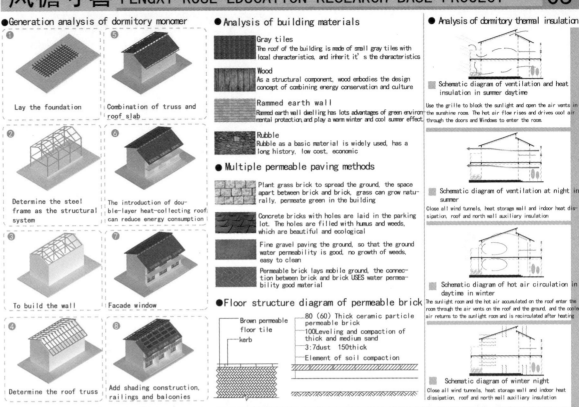

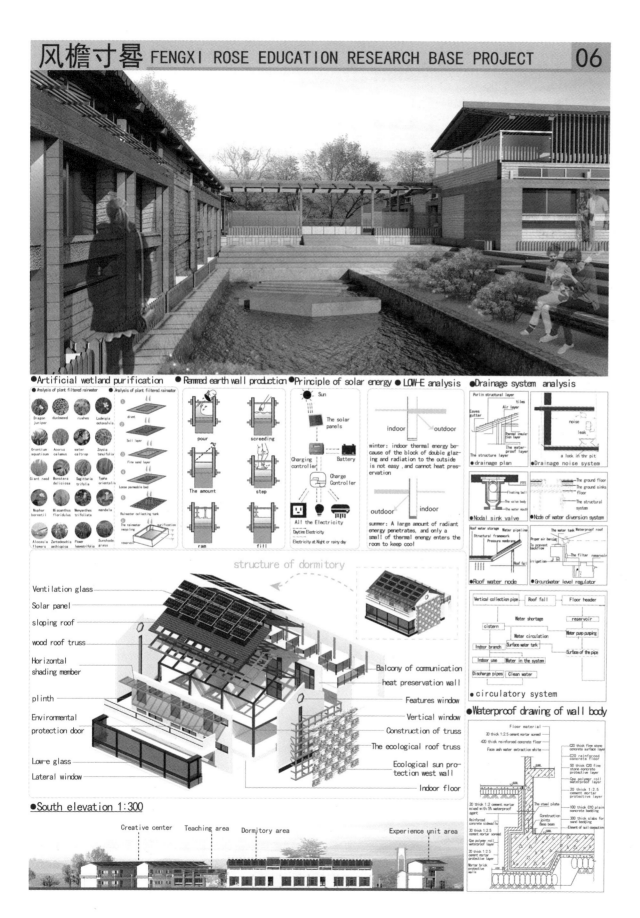

WAITING FOR THE WIND

综合奖·优秀奖
General Prize Awarded·
Honorable Mention Prize

注 册 号：5920
项目名称：等风来（凤溪）
　　　　　Waiting for the Wind (Fengxi)
作　　者：高海伦、方鑫磊、沈令婉、
　　　　　梅博涵、董　兆、季　缘
参赛单位：浙江大学、华南理工大学
指导老师：王国光、王　洁、浦欣成

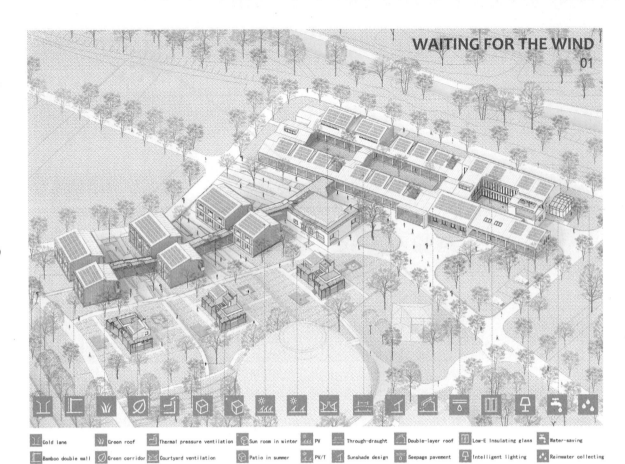

| Cold lane | Green roof | Thermal pressure ventilation | Sun room in winter | PV | Through-draught | Double-layer roof | Low-E Insulating glass | Water-saving |
| Bamboo double wall | Green corridor | Courtyard ventilation | Patio in summer | PV/T | Sunshade design | Seepage pavement | Intelligent lighting | Rainwater collecting |

CHAPTER 1. PRE-PROGRAM

1.1 BACKGROUND ANALYSIS

● SITE ANALYSIS

Tonglu·Hangzhou　Site Position　Surrounding　Field

● TONGLU IMAGE ANALYSIS

富春江 Fuchun River　古村落 Rural　巷道 Hang　竹编 Bamboo

1.2 TRADITIONAL ARCHITECTURAL ANALYSIS

● ENHANCE THE AIR FLOW

Through-draught　Fuzzy space　Patio　Cold lane

● AVOID DIRECT SUNLIGHT

Eaves gallery　Bamboo weaving-double wall　Cornice　Arttic

1.3 CLIMATE ANALYSIS

● SOLAR RADIATION

east　north　south
southeast　southwest　west

From the analysis of different orientations of solar radiation, it is concluded that the eastern, southern and southeastern orientations of solar radiation are higher, so attention should be paid to these orientations of sunshade and sunscreen in architectural design.

● WIND ANALYSIS

Spring　Summer　Autumn　Winter

According to the wind direction and wind frequency analysis of the site, the dominant wind direction in the summer is southeast wind, while the dominant wind direction in winter is north wind. The general plane architectural layout is good for ventilation to form a wind tunnel for cooling and dehumidification while prevent the north wind.

● WIND ENVIRONMENT

Summer

Winter

The site is surrounded by four mountains. Because the dominant wind direction in the summer is southeast wind, the southeast wind passes through the mountains. And the wind speed of the site reach 2.1m/s, which is conducive to summer cooling and dehumidification. The general layout of the buildings should consider the winter wind.

WAITING FOR THE WIND
02

2019 台达杯国际太阳能建筑设计竞赛获奖作品集

Entrance Courtyard

CHAPTER 2: CONCEPT

Zhejiang Tonglu is hot in summer and cold in winter. The buildings need to meet the requirements of heat protection, shading, ventilation and cooling in summer, as well as cold protection in winter. Natural ventilation in passive technology is the main measure to improve the local indoor environment.

Based on the passive energy-saving design, this project takes the patio, the porch and the cold alley of the traditional local dwellings as the prototype. It focuses on the measures of chimney ventilation and draughts. The ventilation design takes advantage of the wind and thermal pressure in high temperature. At the same time, the local bamboo weaving is applied to the exterior wall in order to form a double wall. The project is based on the wind, light and thermal environment simulation to determine the active energy-saving technology for enhancing the thermal performance of the building. The purpose is to make the indoor environment more comfortable. In addition, the project confuses the distinction of interior and exterior by means of the grey space, so as to make the educational research base close to nature.

● DESIGN DESCRIPTION

浙江桐庐属夏热冬冷地区，建筑物需满足夏季防热、遮阳、通风降温要求，冬季应兼顾防寒。被动式技术中的自然通风措施是改善当地室内环境的主要措施。

本方案从被动式节能设计出发，以当地传统民居中的天井、檐廊、冷巷为原型，在此基础上着重考虑拔风、穿堂风等措施，利用当地夏季的高温进行热压、风压共同作用下的通风设计。同时巧妙地将当地常见的竹编用于建筑外墙侧以形成双层墙，基于风、光、热环境模拟来确定主动式节能技术以增强建筑的热工性能，旨在提高室内舒适度。另外方案通过灰空间的设计模糊了室内外，以期形成一个亲近自然的教育研学基地。

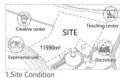
1. Site Condition
The size of the site is 1590m2. It has four functions.

2. Road Setting
Set the roadway and connect with the existing entrance.

3. Functional Division
Divide the site into four parts according to different functions.

4. Courtyard Inside
Place green and public space into each part.

5. Partition Unicom
Connect different parts with corridor system.

6. Roof Change
Use the traditional slope roof, which is more climate-adaptable.

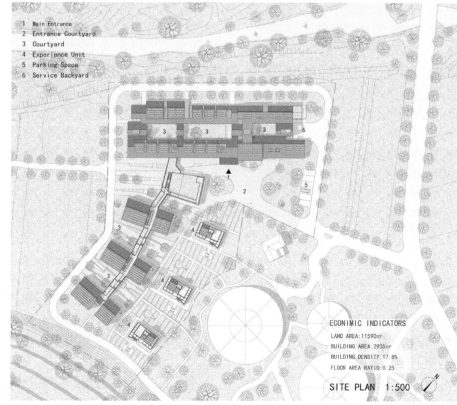

1. Main Entrance
2. Entrance Courtyard
3. Courtyard
4. Experience Unit
5. Parking Space
6. Service Backyard

ECONIMIC INDICATORS
LAND AREA: 11590m²
BUILDING AREA: 2935m²
BUILDING DENSITY: 17.8%
FLOOR AREA RATIO: 0.25

SITE PLAN 1:500

WAITING FOR THE WIND
03

Teaching Center

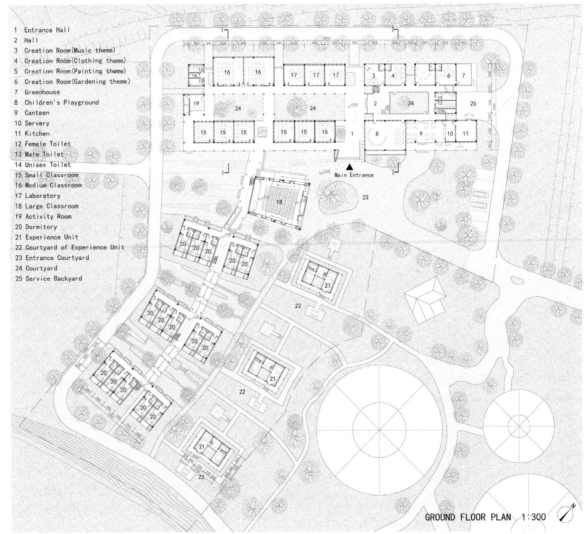

1 Entrance Hall
2 Hall
3 Creation Room (Music theme)
4 Creation Room (Clothing theme)
5 Creation Room (Painting theme)
6 Creation Room (Gardening theme)
7 Greenhouse
8 Children's Playground
9 Canteen
10 Servery
11 Kitchen
12 Female Toilet
13 Male Toilet
14 Unisex Toilet
15 Small Classroom
16 Medium Classroom
17 Laboratory
18 Large Classroom
19 Activity Room
20 Dormitory
21 Experience Unit
22 Courtyard of Experience Unit
23 Entrance Courtyard
24 Courtyard
25 Service Backyard

GROUND FLOOR PLAN 1:300

WAITING FOR THE WIND
04

Creative Center

1 Shared Creation Room
2 Terrace
3 Dormitory
4 Plank Road
5 Experience Unit

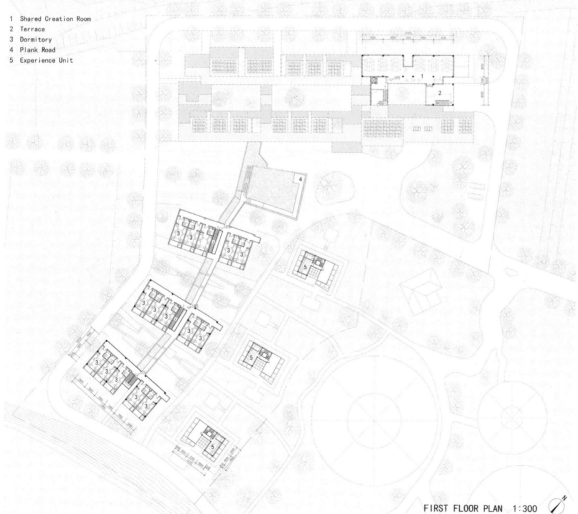

FIRST FLOOR PLAN 1:300

WAITING FOR THE WIND
05

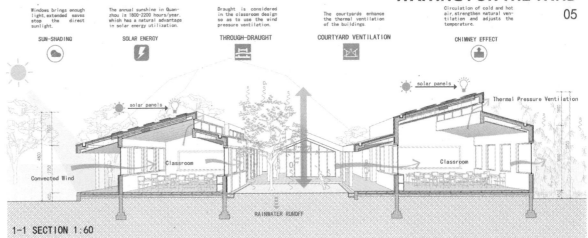

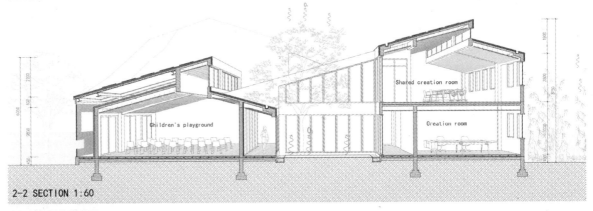

- Dormitory Analysis
- Creative Center and Teaching Center Analysis

CHAPTER 3: GREEN TECHNOLOGY

- Shadow Analysis
- Solar Analysis
- Bamboo Weaving Structure

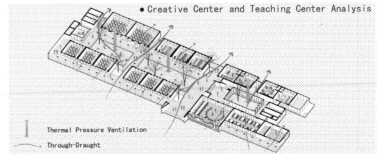

Strengthen Southwest Wind in Summer

Prevent North Wind in Winter

WAITING FOR THE WIND
06

2019 台达杯国际太阳能建筑设计竞赛获奖作品集

- Experience Unit Analysis

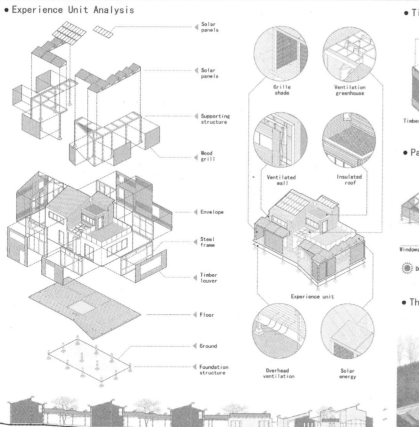

- Timber Louver Structure

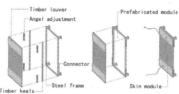

- Passive Energy Saving of the Patio

- Thermal Pressure Ventilation

East Elevation 1:300

South Elevation (Teaching Center and Creative Center \ Dormitory \ Experience unit) 1:300

综合奖·优秀奖
General Prize Awarded · Honorable Mention Prize

注 册 号：5945
项目名称：冷巷·风院（凤溪）
　　　　　Cold Lane · Wind Yard
　　　　　(Fengxi)
作　　者：秦朗、田春来
参赛单位：重庆大学
指导老师：周铁军、张海滨

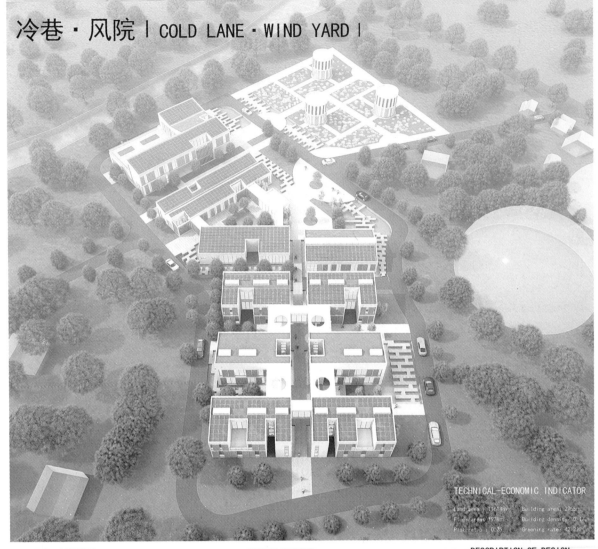

冷巷·风院 | COLD LANE · WIND YARD |

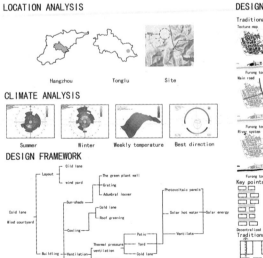

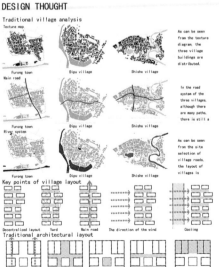

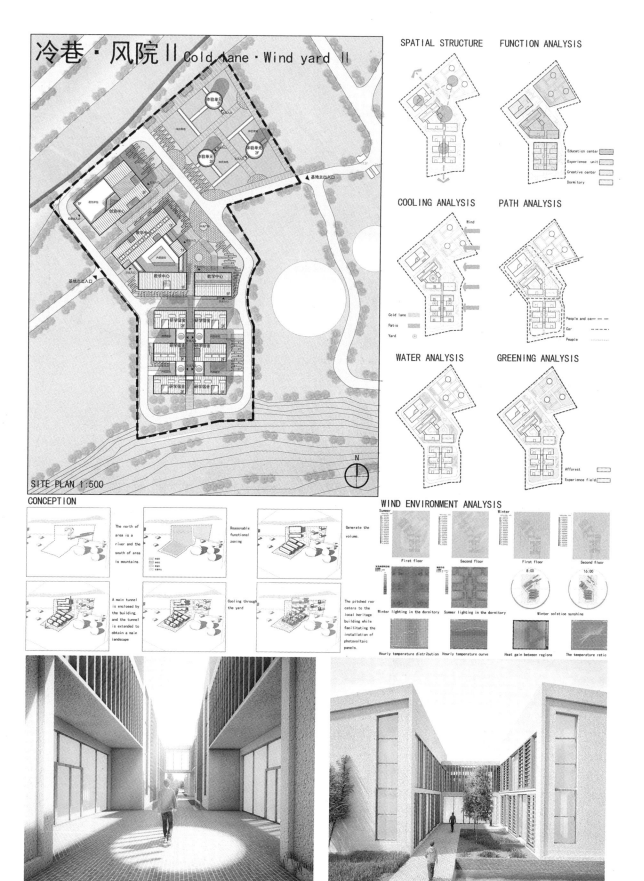

冷巷·风院 III Cold lane · Wind yard III

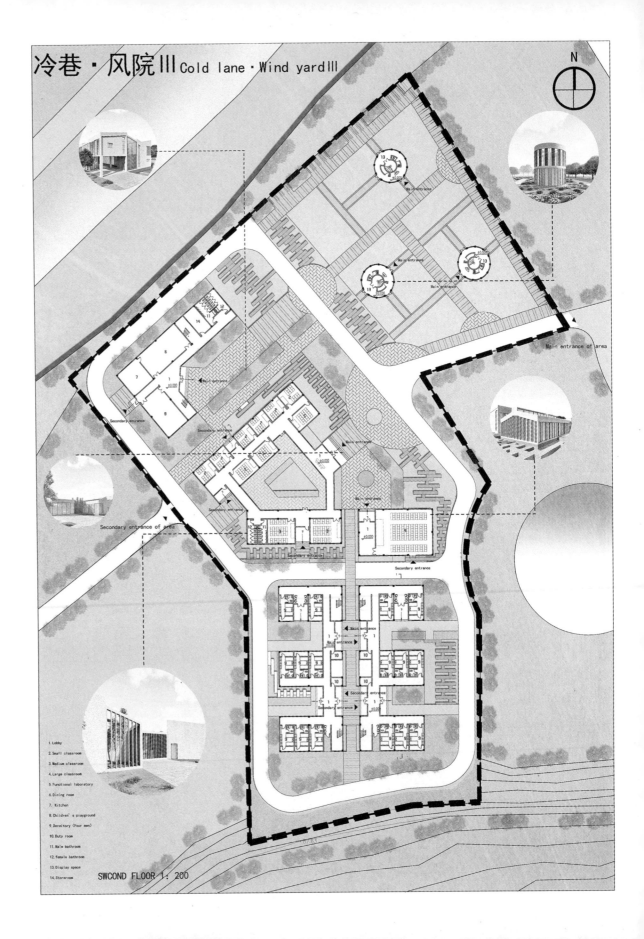

1. Lobby
2. Small classroom
3. Medium classroom
4. Large classroom
5. Functional laboratory
6. Dining room
7. Kitchen
8. Children's playground
9. Dormitory (Four men)
10. Duty room
11. Male bathroom
12. Female bathroom
13. Display space
14. Storeroom

SWCOND FLOOR 1:200

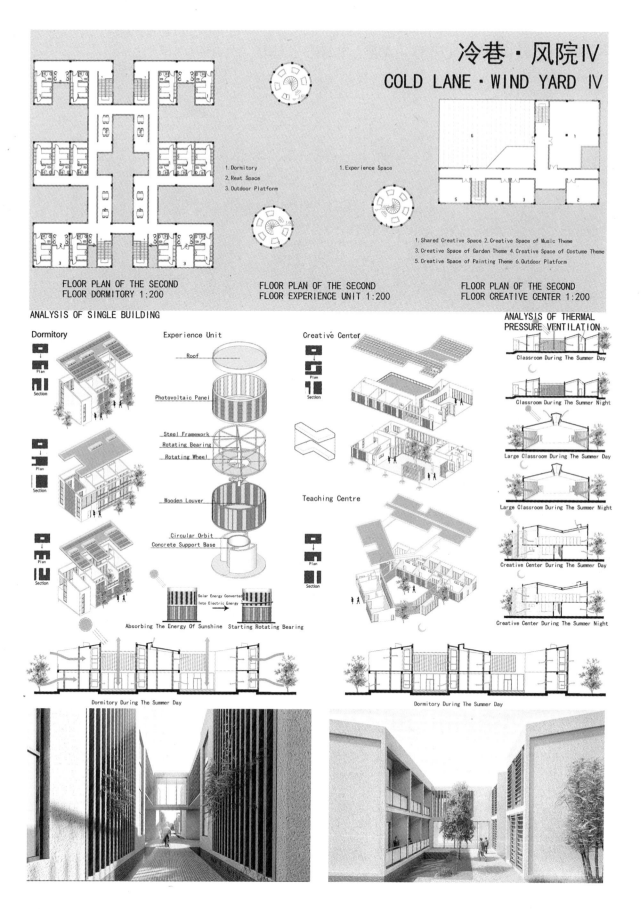

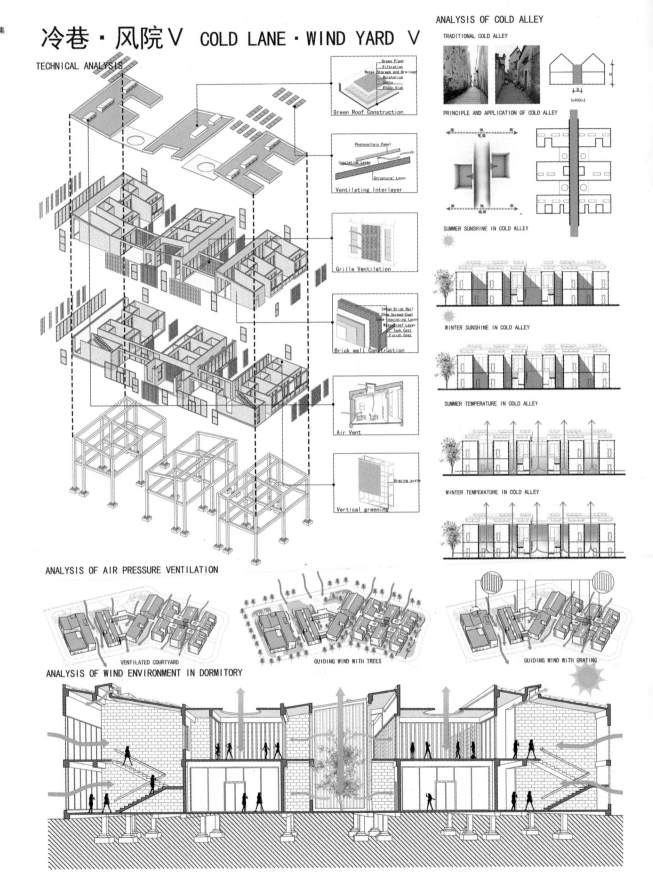

冷巷·风院 VI COLD LANE · WIND YARD VI

1-1 SECTION 1:200   2-2 SECTIONG 1:200

NORTH ELEVATION 1:200

EAST ELEVATION 1:200

WIND ENVIRONMENT ANALYSIS OF CREATIVE CENTER

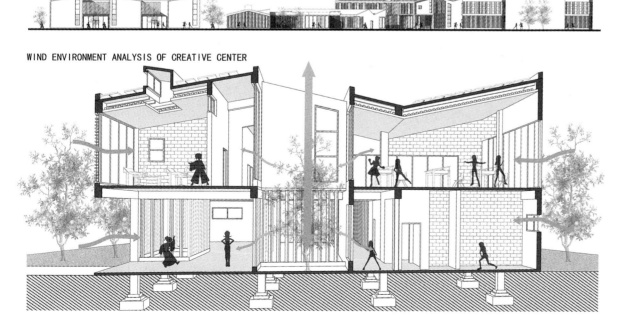

综合奖·优秀奖
General Prize Awarded · Honorable Mention Prize

注 册 号：6009
项目名称：水院风堂（凤溪）
　　　　　Water Courtyard Wind Hall (Fengxi)
作　　者：韩文颢、刘璐瑶
参赛单位：天津大学
指导老师：贾巍杨、王小荣

水院风堂　WATER COURTYARD WIND HALL
ZHEJIANG FENGXI ROSE EDUCATION RESEARCH BASE DESIGN

01

BACKGROUND ANALYSIS
SITE ANALYSIS

SURROUNDING WATER SYSTEM AND BUILDINGS ANALYSIS

Tonglu town, where zhejiang fengxi rose research base is located, is adjacent to the fuchun river with obvious architectural styles Tonglu town, where zhejiang fengxi rose research base is located, is near the fuchun river. There are many scenic spots around the fuchun river, and its southern architectural style is obvious

CLIMATE ANALYSIS

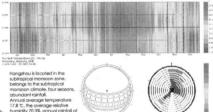

Hangzhou is located in the subtropical monsoon zone, belongs to the subtropical monsoon climate, four seasons, abundant rainfall.
Annual average temperature 17.8 ℃, the average relative humidity 70.3%, annual rainfall of 1454 mm, annual sunshine hours 1765 hours.

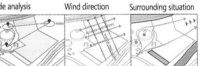
Node analysis　Wind direction　Surrounding situation

ECONOMIC AND TECHNICAL INDICATORS
Building area: 2950 m²;
Floor area: 1710 m²;
Building Density: 11%;
Floor Area thewire: 0.21

SITE PLAN 1:1000

DESIGN DESCRIPTION

桐庐气候温润，不仅有着优美的自然山水，更有着深厚的人文积淀。周边群山中散落着无数自然生长的古村落，其间人与自然和谐互动，潜移默化地形成了多种"绿色技术"（暗渠、明沟、坎井、水塘），不仅构思巧妙，而且易于施工。该方案意图借鉴周边古村的生成逻辑，用古老的智慧呼应这片场地曾经拥有，现在却已经丧失的"文脉"。
在绿色技术上，我们更多选择"被动式太阳能技术"。在整体布局上，该方案综合考虑日照、风向、温度等因素，进行建筑布局，形成三条东西走向"风庭"，疏导夏季过多热量；并于每个北侧设置多层实墙，抵挡冬季强风，形成自身微气候。在建筑单体上，主要通过设置"阳光间"、"通风井"、"遮阳板"等设施，利用内部气流，改善室内环境。

Tonglu has a warm climate, not only with beautiful natural landscapes, but also with profound human accumulation. There are numerous ancient villages scattered in the surrounding mountains. During this period, people and nature interact harmoniously, forming a variety of "green technologies" (ditch, open ditch, and pond), which are not only cleverly conceived, but also easy to construct. The plan intends to draw on the logic of the surrounding ancient villages and use the ancient wisdom to echo the "culture" that the site once owned and now has lost.
In the aspect of green technology, we choose "passive solar technology" more. In the overall layout, the plan takes into account the factors such as sunshine, wind direction and temperature, and carries out the layout of the building, forming three east-west "wind courts" to guide the excessive heat in summer; and setting up multiple solid walls on each north side to withstand the winter. Strong winds form their own microclimate. In the construction unit, the indoor airflow is used to improve the indoor environment mainly by setting up "sunlight", "ventilation well", "sun visor" and other facilities.

水院风堂 WATER COURTYARD WIND HALL
ZHEJIANG FENGXI ROSE EDUCATION RESEARCH BASE DESIGN

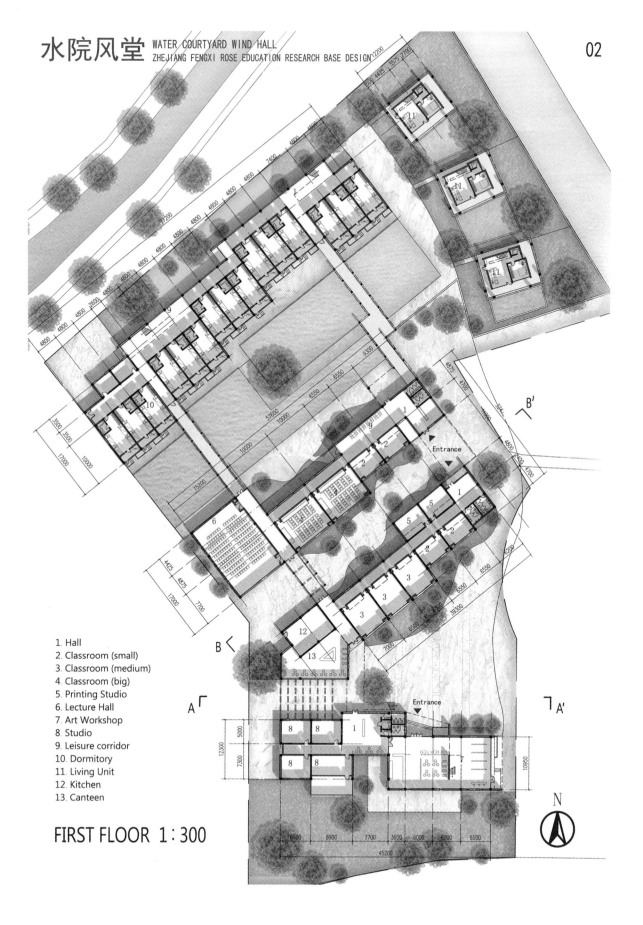

1. Hall
2. Classroom (small)
3. Classroom (medium)
4. Classroom (big)
5. Printing Studio
6. Lecture Hall
7. Art Workshop
8. Studio
9. Leisure corridor
10. Dormitory
11. Living Unit
12. Kitchen
13. Canteen

FIRST FLOOR 1:300

水院风堂 WATER COURTYARD WIND HALL
ZHEJIANG FENGXI ROSE EDUCATION RESEARCH BASE DESIGN

03

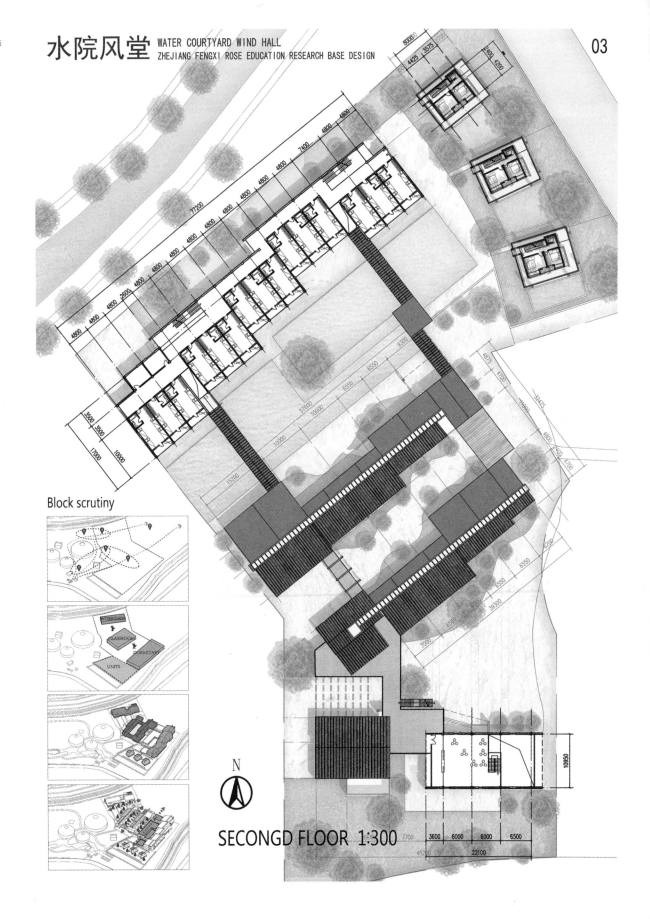

Block scrutiny

SECONGD FLOOR 1:300

水院风堂 WATER COURTYARD WIND HALL
ZHEJIANG FENGXI ROSE EDUCATION RESEARCH BASE DESIGN

04

POINT PERSPECTIVE

ARCHITECTURAL DETAIL

Functional partition

STRUCTURE
Steel frame structures and load-bearing columns conform to architectural design standards

ROOF
The roofs are mostly fiberglass shingles, as well as painted concrete and wood

WALL
The interior features clear zoning and streamline

Passage

Rest

Teach

Active

Entrance to the teaching and residence area

Entrance to the arts centre

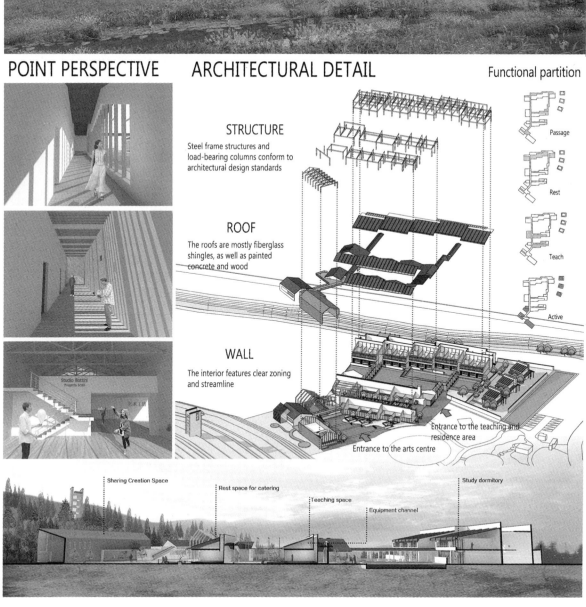

Sharing Creation Space | Rest space for catering | Teaching space | Equipment channel | Study dormitory

水院风堂 WATER COURTYARD WIND HALL
ZHEJIANG FENGXI ROSE EDUCATION RESEARCH BASE DESIGN

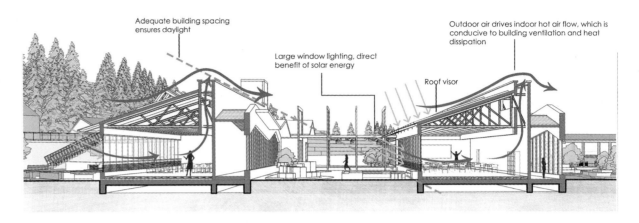

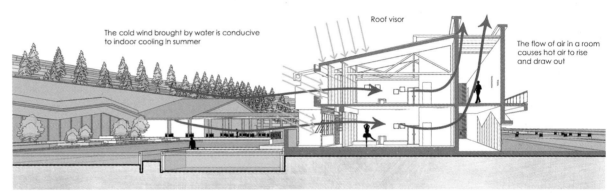

BUILDING MATERIALS ANALYSIS

SOLAR ENERGY APPLICATION

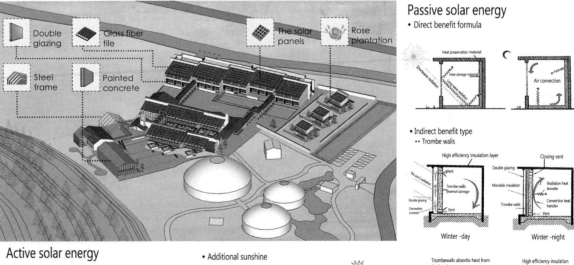

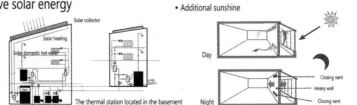

水院风堂 WATER COURTYARD WIND HALL
ZHEJIANG FENGXI ROSE EDUCATION RESEARCH BASE DESIGN

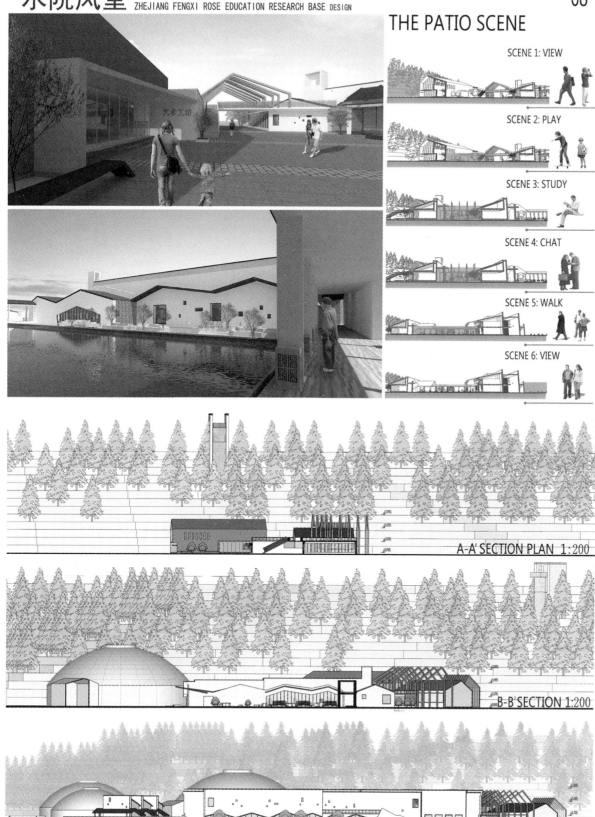

2019台达杯国际太阳能建筑设计竞赛获奖作品集

综合奖·优秀奖
General Prize Awarded·
Honorable Mention Prize

注 册 号：6052
项目名称：檐下风致（凤溪）
　　　　　Leisure Relaxation and
　　　　　Education Under the Roof
　　　　　（Fengxi）
作　　者：苏　珊、王佳璇、叶芷茵
参赛单位：广州大学
指导老师：李　丽、万丰登、夏大为

区位分析 LOCATION ANALYSIS

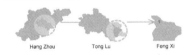

气候分析 CLIMATE ANALYSIS

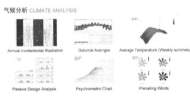

传统建筑原型
TRADITIONAL ARCHITECTURAL
PROTOTYPES

场地风环境模拟 SITE WIND SIMULATION ANALYSIS

日照分析 INSOLATION ANALYSIS

日影分析 SHADOW MAP

建筑技术说明 TECHNICS OF BULIDING DESIGN

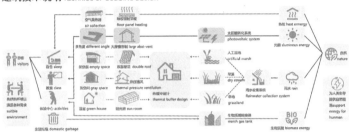

建筑活动说明 ACTIVITIES OF BULIDING DESIGN

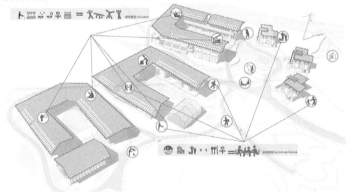

设计说明 DESIGN DESCRIPTION

坡屋顶作为中国南方传统建筑的元素对浙江地域气候具有良好的适应性。

设计通过三个半围合形的大屋檐与围护结构的不同围合方式能形成檐下的走道，架空等空间感受丰富的建筑灰空间，以满足游客各种活动功能的需求。

技术上，从被动式的节能技术出发：通过大屋檐形成的建筑自遮阳体系以及雨水收集系统紧密相连的建筑不同下垫面的庭院形成缓冲区，配合通风屋顶和架空楼板、可移动百叶、采光天窗等，控制建筑人使用空间的温度和采光。在主动太阳能利用上：利用大屋顶结合太阳能光电板、遮阳式新风系统、太阳能空气集热器对太阳能加以运用。

The sloping roof,which is universally acknowledged as a southern traditional architectural element and showing an excellent adaptability towards Zhejiang climate.

From a design perspective,the different enclosure of the half-closed U-type eaves and exterior protected construction generates series of grey areas with the ample sense of space such as the aisles undemeath the roof and the built on stilts for all kinds of the purposes of the guests.

From a technical perspective,the buffered space is composed of the architectural self-sunblocked system and the courtyards of different underlying surfaces that belong to the buildings which are closely interlated with the rain collection sysytem.It mainly takes control of the daylighting and the temperature of the users' living spaces by means of the ventilated window、the overhead floor、removable louver and daylighting skylight.Two things mentioned above are both put into effect in the aspect of thepassive energy-saving technology.

It also use the large roof with solar panels、sun-blocked primary air system and solar air -collections to make the utmost of the solar energy in the aspect of the active solar-energy technology.

檐下风致
LEISURE RELAXATION AND EDUCATION UNDER THE ROOF

2019 台达杯国际太阳能建筑设计竞赛获奖作品集

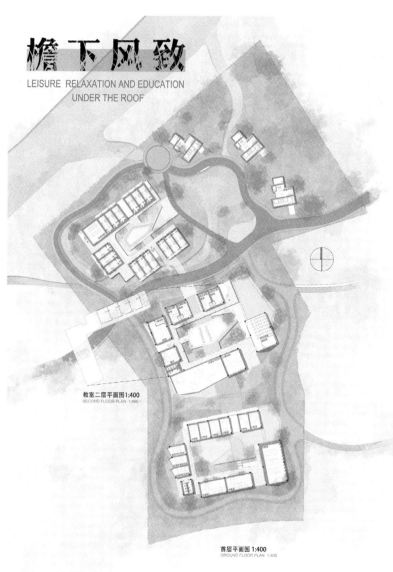

教室二层平面图 1:400
SECOND FLOOR PLAN 1:400

首层平面图 1:400
GROUND FLOOR PLAN 1:400

功能分区
FUNCTION DISTRIBUTION

场地交通分析
TRAFFIC ANALYSIS

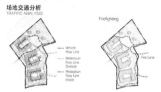

基地交通分析
SITE TRAFFIC ANALYSIS

建筑能耗表
POWER GENERATION OF DORMINTORY UNIT'S SOLAR PANELS

	GENERATED (kWh)	NUMBER (Piece)	AREA (m²)	AVERAGE OUTPUT (KWh/m²)	AVERAGE INPUT (KWh/m²)	AVERAGE EFFICIENCY
YEAR	7304.5	238	215.6	153.5	160	9.9%
SUMMER (7,8)	1256.2	238	215.6	40.3	35	7.5%
WINTER (1,2)	716.5	238	215.6	25.3	25.7	4.3%

宿舍多同组合热辐射对比图
HEAT COMPARISON BETWEEN DIFFERENT COMBINATION

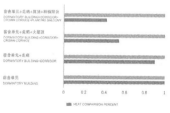

THROUGH THE CALCULATION OF THE DAILY AVERAGE INCIDENT RADIATION OF DIFFERENT COMBINATION DORMINTORY,THE COMPARISON BETWEEN THE FIRST GROUP AND THE SECOND GROUP SHOWS THAT THE AVERAGE DAILY RADIATION IS REDUCED BY 90%,COMPARED WITH THE THIRD GROUP,THE DAILY INCIDENT RADIATION IS REDUCED BY 54% COMPARED WITH THE FOURTH GROUP, THE DAILY INCIDENTRADIATION IS REDUCED BY 42%.

总平面图 1:1000
GENERAL LAYOUT 1:1000

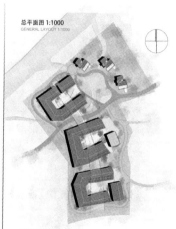

场地风向分析图
SITE WIND DIRECTION ANALYSIS

白天山谷升温快,空气膨胀上升,山谷补充形成谷风
During the day the valley heats up fast, the air expands and the valley supplements to form the Valley wind.

夜晚山坡降温快,空气冷却形成山风
At night, the mountain slopes cool down rapidly and the air cools down to form mountain winds.

Summer — 夏天由于山谷风的影响,场地白天盛行北风
In summer, due to the influence of valley wind, the north wind prevails in the field during the day.

Winter — 冬天受到山谷风的影响,场地盛行西北风
In winter, affected by the valley wind, the northwest wind prevails in the site.

体块分析
PATIO ANALYSIS

设计流程图
DIAGRAM OF DESIGN PROCESS

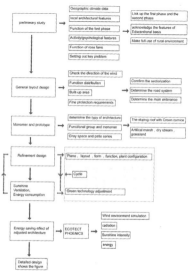

檐下风致
LEISURE RELAXATION AND EDUCATION UNDER THE ROOF

尝试打造一种新的生活模式，培养孩子与自然共生的能力，
启发孩子的理想、希望与意志，让他们为践行自身的梦想走得更远。
We try to create a new life style and cultivate children's ability to coexist with nature, inspire their ideals, hopes and will, and let them go further to practice their dreams.

给孩子们提供亲手触摸微小而美好的事物的机会，
创造一个能拥抱自然、快乐成长的地方。
Give children the opportunity to touch small and beautiful things with their own hands to create a place where they can embrace nature and grow up happily.

追田野里的萤火虫、
光脚在小溪里抓鱼、
躺在石滩上数星星……
这些简单而美好的事物，是孩子最好的启蒙教育。
Chasing the fireflies in the field, barefoot catching fish in the creek, lying on the rocky beach counting the stars …
These simple and beautiful things are the best enlightenment education for children.

植物配置模拟结果 PHOTOSYNTIETICALLY ACTIVE RADIATION

As can be seen from the figure, at the bottom of the eaves of the building, the solar radiation energy is less than 3MJ/m²d, which is suitable for the ombrophyte, such as *Stromanthe sanguinea* Sond. Around the building, the solar radiation energy is between 3MJ/m²d and 5MJ/m²d, which is suitable for planting neutral plants, such as *Rosa cvs*. However, in grassland and roads without building shelter, the solar radiation energy is higher than 5MJ/m²d, which is suitable for the cultivation of heliophilic plants, such as *Oxalis triangularis* and *Sapindus*, etc

植物配置 Garden Plant Layout

西立面图 1：250 West Elevation 1:250

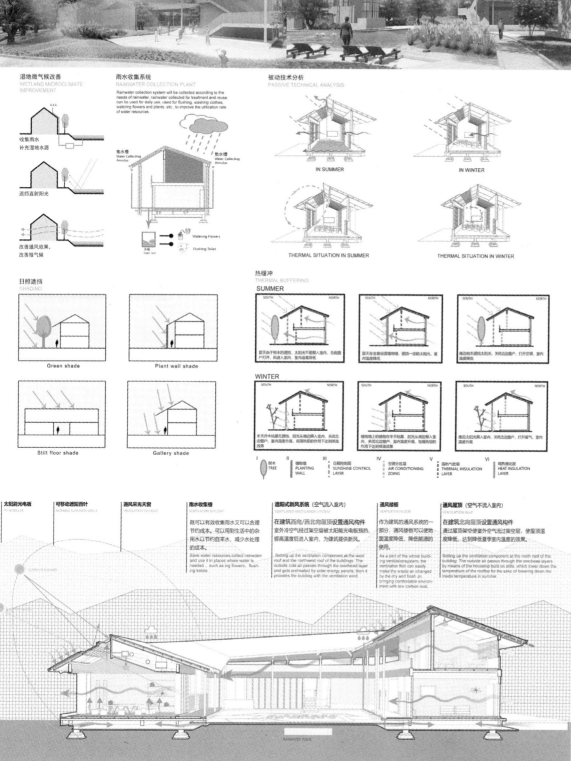

檐下风致

LEISURE RELAXATION AND EDUCATION UNDER THE ROOF

南向教室被动式技术分析
NOTHERN ROOM PASSIVE TECHNICAL ANALYSIS

北向教室被动式技术分析
NOTHERN ROOM PASSIVE TECHNICAL ANALYSIS

西向教室被动式技术分析
WESTERN ROOM PASSIVE TECHNICAL ANALYSIS

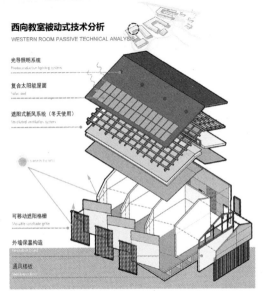

屋顶通风遮阳构件
ROOF VENTILATION AND SHADING ELEMENTS

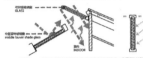

夏季——中置百叶玻璃窗开启至与阳光入射角垂直角度，打开可开闭玻璃窗，阳光从屋顶照射玻璃窗以及竖立的矮墙，导致室内上层温度升高，加上室内有通风楼板，底部温度比上层温度低，形成了向上的风，加强室内通风的效果

In summer the Medium Louver shade glass opens to the vertical angle of the incident with the sun, then opens the glass window next to it, the sun shines on the glass window from the roof and the short wall erected, resulting in the indoor upper temperature rise. Besides there are ventilation floor in the house, the bottom temperature is lower than the upper temperature, which forms the upward wind and strengthens the indoor ventilation effect.

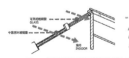

冬季——中置百叶玻璃窗开启至与阳光入射角相同角度，可闭合玻璃窗关闭，阳光从屋顶照射入室内，室内温度上升，从而达到保温的效果

In winter, the Medium Louver shade glass opens to the same angle as the sunny incident angle, then closes the glass window next to it. The sunshine shoots the room from the roof, bring about the indoor upper temperature to rise, the indoor temperature rises, thus achieves the heat preservation effect.

太阳能光伏发电板的即发即用
USE OF SOLAR PHOTOVOLTAIC PANELS

建筑屋顶倾斜角度不同，太阳能光电板放置的角度不同，可以更好的将不同时刻的太阳光转化为电能，对耗电较低的构件（例如导风扇）进行直接供电，省去接入集中电网的耗电量，更好的达到节能效果。

The different angle of the inclination of the building roof and the different angle of the solar-energy panels placement much more properly in every moment. It also supplies the electric power with the low-power component (like ventilation fan) to leave out the energy loss of the central power grid for the sake of the energy conservation.

通风屋顶（空气不流入室内）
VENTILATION ROOF

在建筑北向屋顶设置通风构件
通过顶部架空使室外空气流过空气层，使屋顶温度降低，达到降低夏季室内温度的效果。

Setting up the ventilation component at the north roof of the building. The outside air passes through the overhead layers by means of the housetop build on stilts, which lower down the temperature of the rooftop for the sake of lowering down the inside temperature in summer.

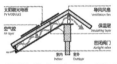

IN SUMMER——When the valve is on, air can get inside and reduce the temperature of insulating layer.
IN WINTER——When the valve is off, air can not pass through the air-filled cavity which lead to roof insulation.

导风风扇的接电方式
POWER CONNECTION MODE OF VENTILATION FAN

通风楼板
VENTILATION FLOOR

通过架空楼板，使室外空气在建筑底部经过，不仅能防潮、防湿，而且能降低室内地面的温度，从而实现室内空气的对流。

结构通风
Structural ventilation

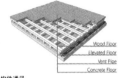

构件通风
Component ventilation

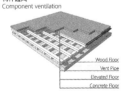

遮阳式新风系统（空气流入室内）
VENTILATED VENTILATION SYSTEM

在建筑西向/西北向屋顶设置通风构件
室外冷空气经过架空层被太阳能光电板预热，提高温度后进入室内，为建筑提供新风。

Setting up the ventilation component at the west roof and the north-west roof of the buildings. The outside cold air passes through the overhead layer and gets preheated by solar-energy panels, than it provides the building with the ventilation wind. When it gets inside with the higher temperature.

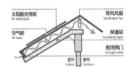

IN SUMMER——When the valve is off, air can not pass through the air-filled cavity.
IN WINTER——When the valve is on, air can get inside after it becomes preheated by photovoltaic panels which are set up at the air-filled cavity.

牵引式垂直绿化结构剖面及局部处理
THE STRUCTURE PROFILE MAP AND THE LOCAL PROCESSING OF TRACTION VERTICAL GREENING

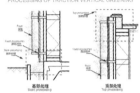

基部处理 / 顶部处理
Based processing / Top processing

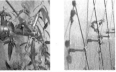

固定附件
Fixed Accessories

檐下风致 阳光·文化之旅
LEISURE RELAXATION AND EDUCATION UNDER THE ROOF
浙江凤溪玫瑰教育研学基地建筑设计

2019 台达杯国际太阳能建筑设计竞赛获奖作品集

宿舍部分节能技术分解
GREEN TECHNICS OF DORMITORY

宿舍单体轴测图
DORMITORY UNIT AXONOMETRIC VIEW

1. 太阳能空气集热器 Solar collector
2. 双光栅结构薄膜太阳能电池 Optimization of thin film solar cells with double grating structure
3. 通风屋面 Ventilation roof
4. 相变储能墙 Phase change wall
5. 单面钢丝网架发泡聚苯乙烯外挂保温板 (DGJ)
6. 架空楼板 Aerial ventilation floor
7. 雨水收集槽 Rainwater collection
8. 牵引测壁式垂直绿化 the attachment of vertical greenning
9. 绿色化种植阳台 Planting balcony

五组宿舍单元+走廊
Five roup of dormitory building +Corridor

五组宿舍单元+走廊+大屋顶
Five roup of dormitory building+Corridor+Crown cornice

五组宿舍单元+走廊+大屋顶+种植阳台
Five roup of dormitory building +Corridor+Crown cornice+Planting balcony

Through the calculation of the daily average incident radiation of different combination dormitory, the comparison between the first group and the second group shows that the average daily incident radiation is reduced by 40%. Compared with the third group, the daily average incident radiation Reduce by 15%.
It can be concluded that the Crown cornice and Planting balconies have a significant reduction in the heat entering the dormitory unit.

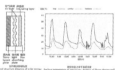

复合相变墙技术详图
TECHNICAL DETAILS OF PHASE CHANGE WALL

选择在室内舒适温度范围内具有高潜热和合适相变温度的呼合比例，用相变储热，白天由太阳能空气集热器加入室内空气，热空气通过相变墙进入室内，相变材料蓄热。夜间相变辐射换热加热室内空气，相变材料放热。

The transition temperature and the heat of the different proportion of the Cpariacacid and Lauric acid mixture are tested by differential sccanning calorimetry, choosing in indoor comfortable temperature range mixture ratip with high latent heat and suitble phase temperature and using phase change material release.

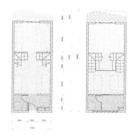

宿舍单元1/2平面1:200　宿舍单元3m平面1:200
宿舍居住单元平面图 1:200
DORMITORY UNIT PLAN 1:200

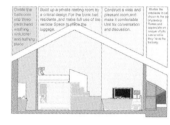

宿舍居住单元立面示意图
DORMITORY UNIT ELEVATION

体验中心节能技术分解
GREEN TECHNICS OF EXPERIENCE CENTER

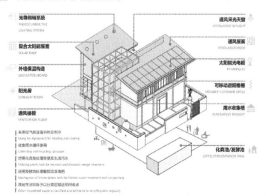
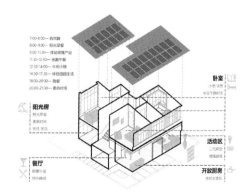

105

综合奖·优秀奖
General Prize Awarded · Honorable Mention Prize

注 册 号：6097
项目名称：研杭光社（凤溪）
Hangzhou Light & Sight Club (Fengxi)
作　　者：刘家韦华、张玉琪、郭嘉钰、胡紫雯
参赛单位：天津大学、河北工程大学
指导老师：田　芳、侯万钧

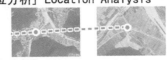

研杭光社 Hang Zhou Light & Sight Club
——凤溪研学旅行基地设计

[区位分析] Location Analysis

[设计说明] Design Description

研杭光社，为研学旅行提供暖如阳光的居所。本方案将杭州传统建筑元素与先进技术结合，并且充分考虑了独特的地理环境，使建筑既与环境契合又成为区域内的亮点。建筑尽可能利用了太阳能资源，并且设计了不同天气的应对措施。同时为学生及其家庭提供了一个亲近自然环境、了解传统文化的活动空间。

This project combines the traditional architectural elements and advanced technology of Hangzhou, and fully considers the unique geographical environment, so that the building not only fits the environment, but also is called the highlight in the region. The building uses solar energy as much as possible, and designs different weather response measures. This program provides a space for children and their families to get close to the natural environment and understand traditional culture.

[气象分析] Climate analysis

Best buliding direction is 202.5 degrees.

[地方特色] Local Characteristics

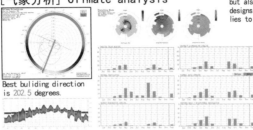

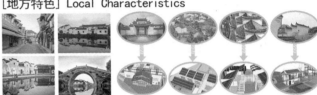

[技术总览] Application of Green Building Technology

- Photovoltaic curtain wall
- Three dimensional greening
- Low-E
- Photovoltaic
- Ventilated roof
- Dehumidifier
- Ventilated roof
- Radiant heating
- Regenerative floor
- Roof Garden
- Active sunshade
- Rainwater recovery
- Thermal insulation wall
- TK chair system

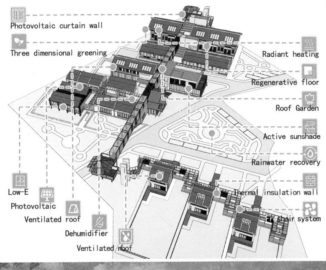

[生成推演] Formation Analysis

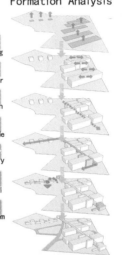

研杭光社 Hang Zhou Light & Sight Club ②
——凤溪研学旅行基地设计

[总平面图] Site Plan 1:1000

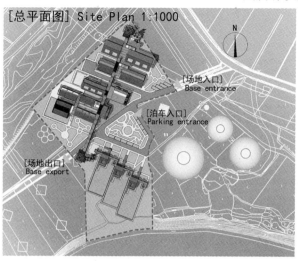

- [场地入口] Base entrance
- [泊车入口] Parking entrance
- [场地出口] Base export

[建筑日影分析] Architectural Sun Shadow

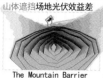

山体遮挡场地光伏效益差
The Mountain Barrier

It is found that the north side of the site is easily covered by the mountain, which is not conducive to the use of solar energy. Therefore, the main building is located in the south side of the site.

Winter Soistice

[技术经济指标] Economic Technological Index

Urban Planning Area:	16142.28㎡
Overall Floorage:	3175.43㎡
Green Rate:	63.13%
Building Density:	0.24
Volume rate:	0.36

Summer Soistice

[场地风环境] Site Wind Environment

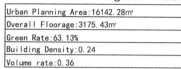

Summer (Day) | Winter (Day)
Summer (Night) | Winter (Night)

[采光模拟] lighting simulation

[体验单元部分节能技术分解] Decomposition of Energy-saving Technology

[拆解前] Before dismantling

1. [复合太阳能屋面] Composite solar roof
2. [通风屋面] Ventilation roof
3. [辐射制冷系统] Radiation refrigeration
4. [外挂保温模塑聚苯板] Insulation board
5. [活动遮阳装置] Movable shading device
6. [辐射供暖系统] Radiation heating system
7. [雨水收集] Rain water collection
8. [立体绿化] Tridimensional virescence

[地源热泵] Ground source heat pump

Flat panel solar energy — Air conditioner — Radiant floor heating — Hot water pump — Indoor water point

Solar Energy Workstation — Buried pipe — GSHP

冬季，地面能量的热量被抽走用于室内采暖；夏季，室内的热量被带走，释放到地下。
In winter, the heat in the ground energy is taken out for indoor heating; in summer, the heat in the room is taken out and released into groundwater, soil or surface water.

Three-dimensional greening | PV panel

Three-dimensional greening decorates the facade and isolates thermal radiation.

Photovoltaic panels best Angle for 22°

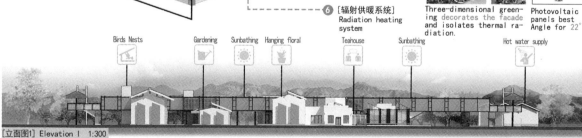

Birds Nests | Gardening | Sunbathing | Hanging floral | Teahouse | Sunbathing | Hot water supply

[立面图1] Elevation I 1:300

Hot water supply | Sunbathing | Teahouse | Hospital | Gardening | Hanging flora | Birds Nests

[立面图2] Elevation II 1:300

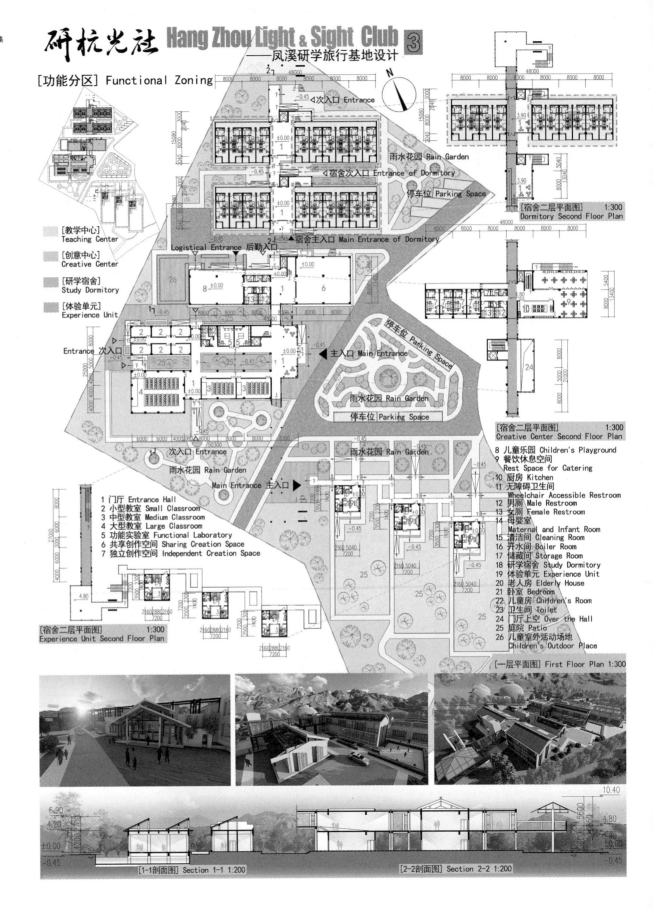

研杭光社 Hang Zhou Light & Sight Club ④
——凤溪研学旅行基地设计

[彩虹除湿装置] Rainbow Dehumidifier

幕墙上设有三棱形玻璃盒，内设干湿机。当除湿器变为液体时，它会形成一个三角棱镜，这会导致色散，并在室内形成彩虹投影，营造愉快的儿童活动空间。

Triprism-shaped glass boxes are arranged on the curtain wall and dry dehumidifiers are placed in them. When the dehumidifier becomes liquid, it forms a triangular prism, which can cause dispersion and form a rainbow projection in the room.

[剖面策略] Section Strategy

[白天] Day　[夜晚] Night
[夏季] Summer　[冬季] Winter
[夏季] Summer　[冬季] Winter

[流水玻璃幕墙] Flowing glass curtain wall

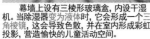

玻璃幕墙两端增设储水池。净化后的雨水引入池中，水流靠近外玻璃的内壁，形成瀑布景观。水可以循环利用。流动式幕墙能有效带走幕墙的热量。

Water storage pools are added at both ends of the glass curtain wall. The purified rainwater is introduced into the pool and the water flows down close to the inner wall of the outer glass to form a waterfall landscape. The water can be recycled. The flow curtain wall can effectively take away the heat of the curtain wall.

[雨水回收利用] Rainwater Recovery and Utilization

[剖透视图] Sectional View 1:200

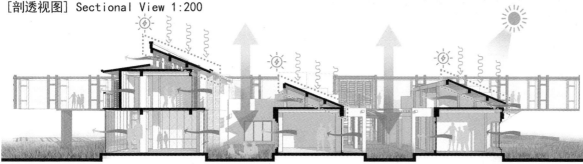

游廊·穿院
浙江凤溪玫瑰教育研学基地设计

01

综合奖·优秀奖
General Prize Awarded · Honorable Mention Prize

注 册 号：6121
项目名称：游廊·穿院（凤溪）
　　　　　Veranda · Courtyard (Fengxi)
作　　者：王少杰、丁　昺、柳天元
参赛单位：沈阳建筑大学、
　　　　　丝路视觉科技股份有限公司
指导老师：王常伟、汝军红、王　飒

本次设计基于中国传统庭院及新中式建筑形式，以充分利用太阳能等可再生能源技术为主要目的，通过对建筑与场地进行推敲，使得建筑用地能够得到最大化的利用。
设计以院落为主题对建筑体块进行演变，使得建筑与庭院融为一体，增加了室内与室外的联系。同时对现代建筑的元素进行提取与使用，并结合现代光伏技术与外围护节能技术，从而打造出绿色节能的教学中心。

The design is based on the traditional Chinese courtyards and the new Chinese style of architecture, and takes full advantage of the renewable energy technologies such as solar energy as the main purpose.
Design to the courtyard as the theme of the evolution of the building block, making the building and courtyard integration. Increased indoor and outdoor links. At the same time, the elements of modern architecture are extracted and used. Combined with modern photovoltaic technology and exterior protection energy-saving technology, a green and energy-saving teaching center can be created.

CHAPTER 01: PRE-PROGRAM

1.1 [Climate Analysis]

Solar Radiation　　Orientation　　Psychrometric Chart　　Prevailing Winds　　Weekly Summary

1.2 [Location Analysis]

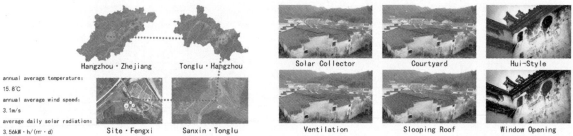

1.3 [Site Analysis]

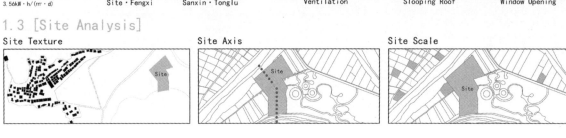

Site Texture　　Site Axis　　Site Scale

游廊·穿院　浙江凤溪玫瑰教育研学基地设计

Techno-economic Indicator
Land Area: 11611.8m²
Building Area: 2912.5m²
  First Floor: 1849.3m²
  Second Floor: 1063.2m²
Building Density: 25.1%
Floor Area Ratio: 0.16

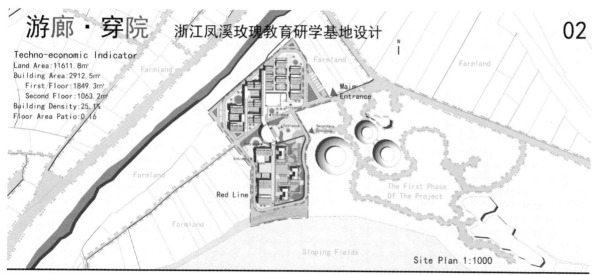

Site Plan 1:1000

CHAPTER 02: PROJECT DESIGN

1-Site
2-Function (1-Classroom, 2-Dormitory, 3-Experience, 4-Creation Space)
3-Subdivision
4-Detailing
5-Courtyard
6-Corridor
7-Slooping Roof
8-Surrounding
9-Framework

Courtyard

Framework

Liana · Sunlight Grennhouse · Water System · Solar Collector · Skylight · Roof Planting · Sunlight Grid · PV Modules · Courtyard

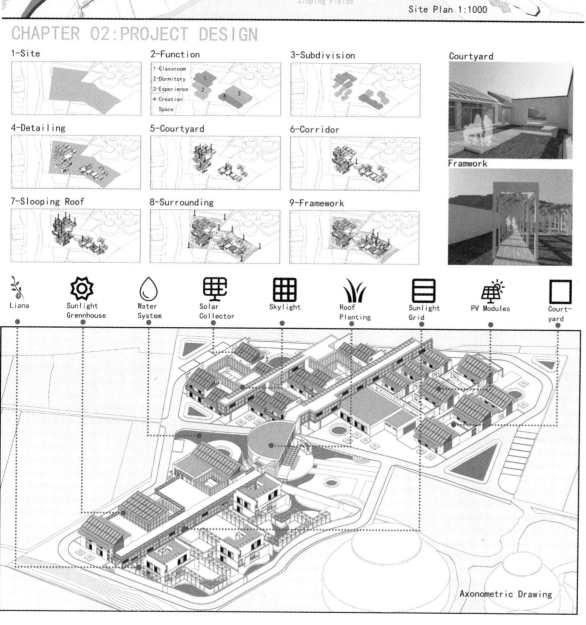

Axonometric Drawing

游廊·穿院　浙江凤溪玫瑰教育研学基地设计

03

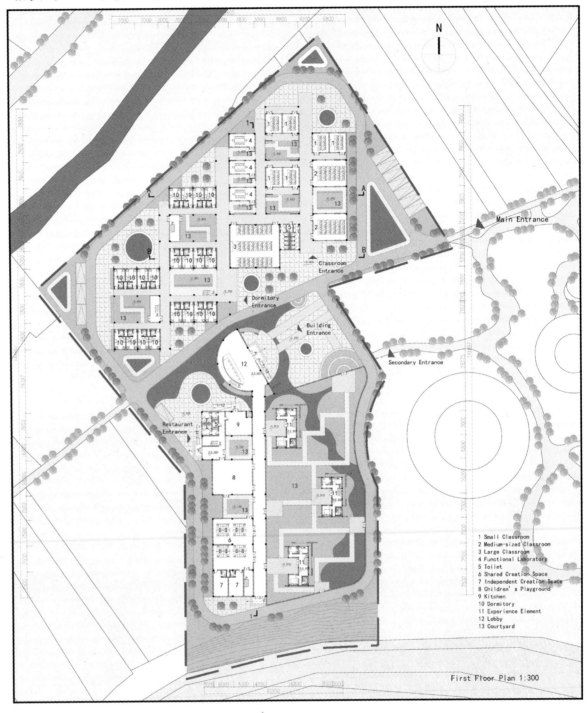

游廊·穿院　浙江凤溪玫瑰教育研学基地设计

04

1 Independent Creation Space
2 Dining And Rest Space
3 Dormitory
4 Experience Element
5 Void
6 Roof
7 Sunlight Greenhouse

Second Floor Plan 1:300

North Elevation 1:300

West Elevation 1:300

Sunlight Greenhouse　Lobby　Courtyard　Corridor

游廊·穿院 浙江凤溪玫瑰教育研学基地设计

CHAPTER 03: PROJECT TATIC

3.1 [Energy Conservation Technique]

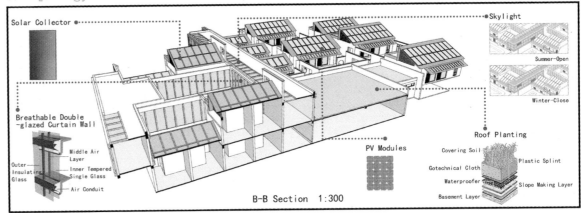

3.2 [Enlarged Drawing——Dormitory 1:50]

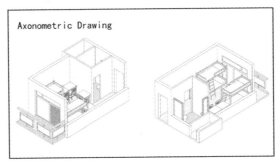
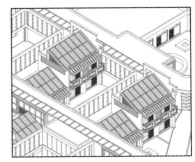

3.3 [Ventilating System]

3.3.1 Summer
3.3.2 Winter

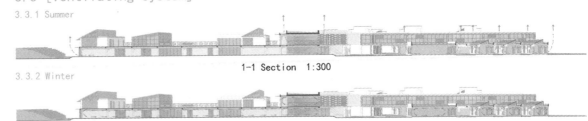

3.4 [Recycled Water System]

游廊·穿院 浙江凤溪玫瑰教育研学基地设计

06

3.5 [Exploded Drawing]

3.5.1 Small Classroom

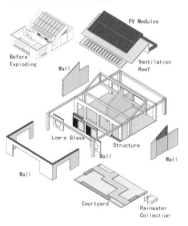

3.5.2 Dormitory

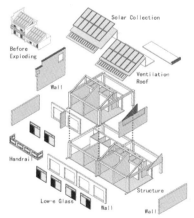

3.5.3 Experience Element

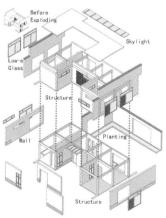

3.6 [Sunlight Analysis]

3.6.1 Shadow Analysis

Spring Equinox

Summer Solstics

Autumn Equinox

Winter Solstics

3.6.2 Sunlight Analysis

Winter Solstics

综合奖·优秀奖
General Prize Awarded · Honorable Mention Prize

注 册 号：6252
项目名称：光伏森林（凤溪）
　　　　　Forest of Solar Photovoltaics
　　　　　(Fengxi)
作　　者：仲文洲、刘　巧、隋明明、
　　　　　李心怡
参赛单位：东南大学、东京工业大学
指导老师：张　彤

Forest of Solar photovoltaics

Construction Area	2902.81 ㎡
Site Area	11801.23 ㎡
Floor Area	2101.29 ㎡
Floor Area Ratio	0.24
Building Density	0.23
Greening Rate	70.60%

设计说明

立足场地自然要素，结合在地气候条件，利用并强化原有生态资源发展农业景观，合理布局建筑与场所营造，最优化建筑风、光、热环境，创造舒适可持续的场地环境；以浙江民居作为被动式设计原型，拓展以"合院"、"冷巷"为核心的传统气候调节手段，辅以热质技术特朗伯墙（Trombe Wall）以弥补前者冬季采暖的不足；以植物"光合作用"作为主动式设计原型构筑光伏森林，为建筑供给能量、遮阳，并营造舒适宜人的游廊空间；从总体布局到建筑细部注重主被动技术结合，合理利用乡土建构的材料与工法，利用夯土、再生竹木与在地山石，减少碳排放，重拾文化特征。

Based on the natural elements, climatic conditions and ecological resources, the building layout is developed rationally to optimize the wind, light and thermal environment, creating a comfortable and sustainable site;
We expand the traditional passive technologies applied in vernacular buildings of Zhejiang Province as a passive design prototype. Thermal mass technologies like Trombe Wall are supplemented to make up for the lack of winter heating.
Take the photosynthesis of plants as active design prototype, a photovoltaic forest will be built to provide energy and shading for the building, also create comfortable and pleasant veranda space.
From the overall layout to the architectural details, we focus on the combination of active and passive technologies, rational use of materials and methods of local construction, including the agricultural planting, use of bauxite, recycling bamboo and wood, which would reduce carbon emissions and regain cultural identity.

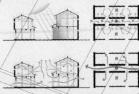

Environmental regulation by courtyard & corridor

How to organically unit natural ecology, residence dwelling and photovoltaic systems?

Site plan 1:600

| Best orientation for solar photovoltaics | Grid change based on terrain and wind | Building connection | Forest of solar photovoltaics |

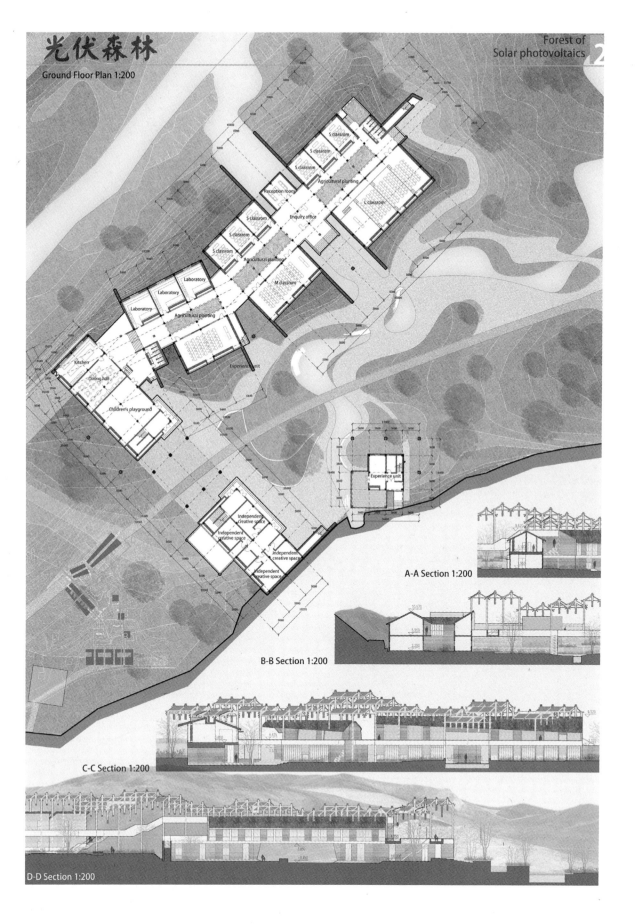

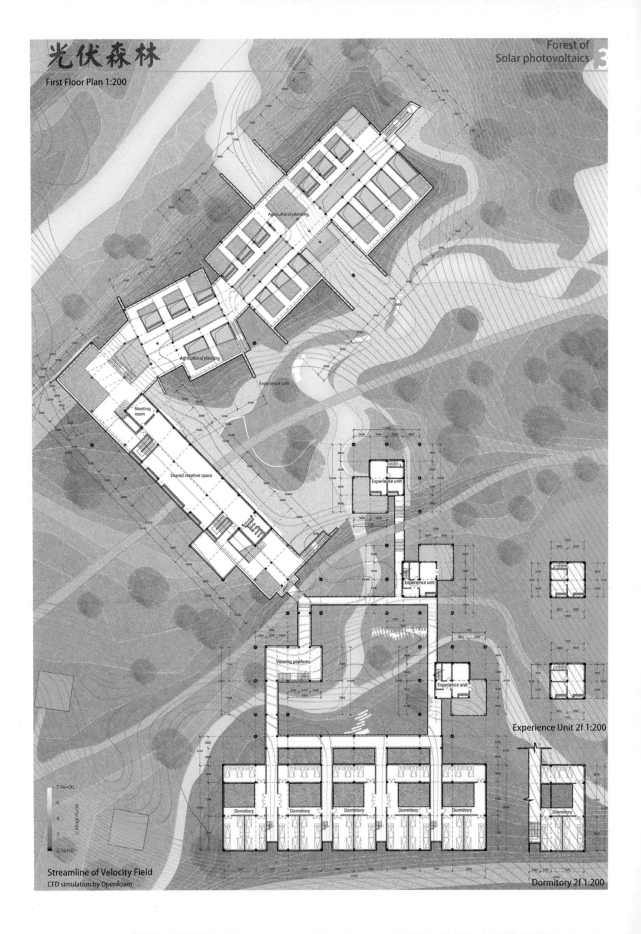

光伏森林 / Forest of Solar photovoltaics

光伏森林 / Forest of Solar photovoltaics

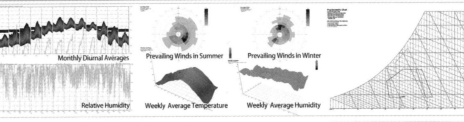

Climate Analysis

According to the Ecotect analysis on local climate, several passive technologies can be applied to the program, which would benefit the physical environment to a large extent. They are:

- Thermal mass effects
- Exposed mass + night-purge
- ventilation
- Natural ventilation
- Passive solar heating

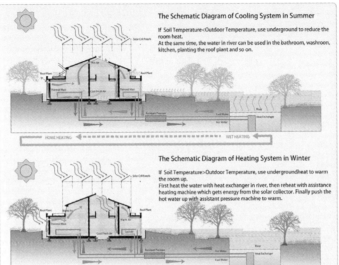

The Schematic Diagram of Cooling System in Summer

If Soil Temperature<Outdoor Temperature, use underground to reduce the room heat.
At the same time, the water in river can be used in the bathroom, washroon, kitchen, planting the roof plant and so on.

The Schematic Diagram of Heating System in Winter

If Soil Temperature>Outdoor Temperature, use undergroundheat to warm the room up.
First heat the water with heat exchanger in river, then reheat with assistance heating machine which gets energy from the solar collector. Finally push the hot water up with assistant pressure machine to warm.

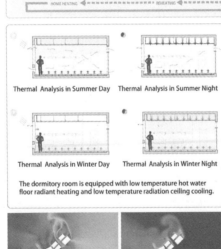

Thermal Analysis in Summer Day | Thermal Analysis in Summer Night
Thermal Analysis in Winter Day | Thermal Analysis in Winter Night

The dormitory room is equipped with low temperature hot water floor radiant heating and low temperature radiation ceiling cooling.

Materail and energy cycle of argricultural landscape,domestic garbage and aquculture.

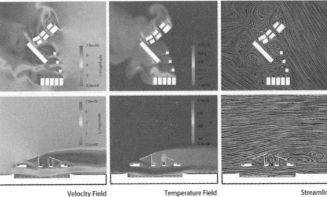

Velocity Field | Temperature Field | Streamline

Validation of natural ventilation by 3D RANS CFD simulations operated on Openfoam

光伏森林

Forest of Solar photovoltaics

综合奖·优秀奖
General Prize Awarded · Honorable Mention Prize

注 册 号：6289
项目名称：云山·风居——浙江凤溪玫瑰教育研学基地设计（凤溪）
Wind House—Design of Rose Education and Research Base in Fengxi, Zhejiang Province（Fengxi）
作　　者：康永基、王润丰、王子祺、郭雪婷
参赛单位：天津大学、河北工业大学
指导老师：戴　路、汪江华、舒　平

Local Traditional Analysis

Adopts the texture of traditional architecture

The traditional roof form is adopted

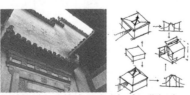
Absorbing the ventilation principle of traditional buildings

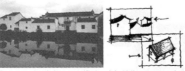
The same materials as traditional buildings are used.

Site Location Analysis

Zhenjiang　　Hangzhou

The project is located in the Fengxi Rose Education Research Study Base in Sanxin Village, Hangzhou City. Tonglu County boasts a subtropical monsoon climate with four distinct seasons, ample sunshine and abundant rainfall. There exists a sound coordination among light, temperature and rainfall, all of which maintain basic synchronization in the increase and decrease, providing abundant climate resources.

Climate Analysis

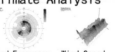
Wind Frequency　　Wind Speed　　Climate Monthly Data

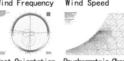
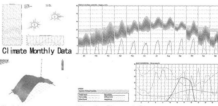
Best Orientation　　Psychrometric Chart　　Weekly temperature　　Hourly Thermal Data

Design Explanation

本方案以凤溪当地传统民居为出发点，将传统建筑形式与绿色建筑技术相融合，以现代形式再现传统文化。建筑以小体量分散布置在场地中，在迎合周边建筑肌理的同时促进了场地通风。建筑汲取了传统的建筑形式而采用坡屋顶，便于充分利用太阳能。建筑中采用了太阳能烟囱、立体绿化、地道风通风系统等绿色建筑技术，为人类创造了健康、舒适、绿色的建筑空间。

This project takes Fengxi local traditional dwellings as the starting point, combines traditional architectural forms with green building technology, and reproduces traditional culture in modern form. The buildings are scattered in small volume in the site, which caters to the texture of the surrounding buildings and promotes the ventilation of the site. The building absorbs the traditional building form and adopts the sloping roof, which is convenient to make full use of solar energy. Green building technologies, such as solar chimney, three-dimensional greening and tunnel ventilation system, are adopted in the building, creating healthy, comfortable and green building space for people.

云山·风居
WIND HOUSE — Design of Rose Education and Research Base in Fengxi, Zhejiang Province

2019 台达杯国际太阳能建筑设计竞赛获奖作品集

Economic and Technical Indicators
- Base area: 11590 m²
- Total building area: 2900 m²
- Volume ratio: 0.25
- Building density: 0.14
- Greening rate: 75%

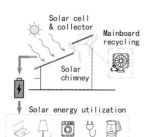

The combination of waste recycling materials and solar photovoltaic panels forms integrated photovoltaic and photoelectric components, making full use of the local high-quality solar energy resources in Fengxi.

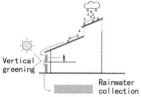

The rain water through the sponge city for the evolution of storage making full use of the local humid and rainy climate conditions.

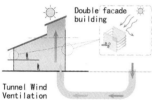

Use the height of the pitched roof to make the height of human activities have a comfortable temperature

Technology Utilization

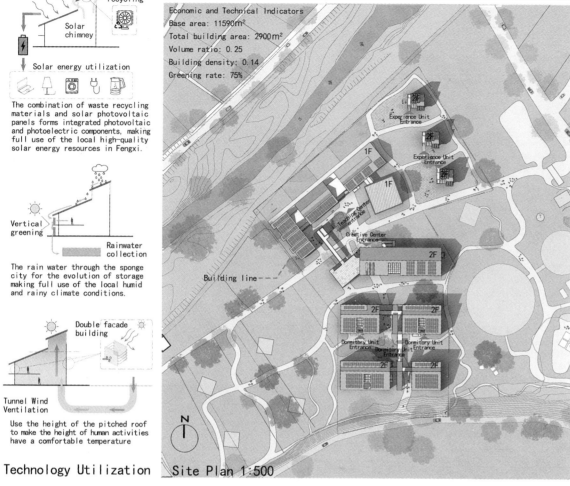

Site Plan 1:500

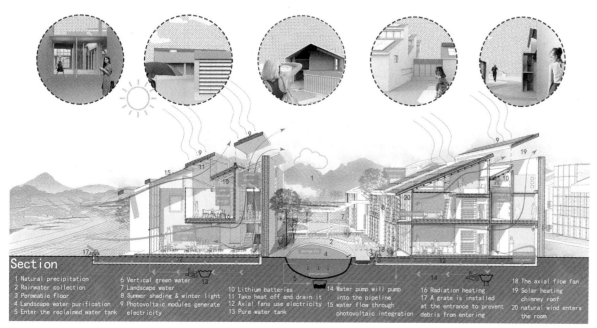

Section
1. Natural precipitation
2. Rainwater collection
3. Permeable floor
4. Landscape water purification
5. Enter the reclaimed water tank
6. Vertical green water
7. Landscape water
8. Summer shading & winter light
9. Photovoltaic modules generate electricity
10. Lithium batteries
11. Take heat off and drain it
12. Axial fans use electricity
13. Pure water tank
14. Water pump will pump into the pipeline
15. water flow through photovoltaic integration
16. Radiation heating
17. A grate is installed at the entrance to prevent debris from entering
18. The axial flow fan
19. Solar heating chimney roof
20. natural wind enters the room

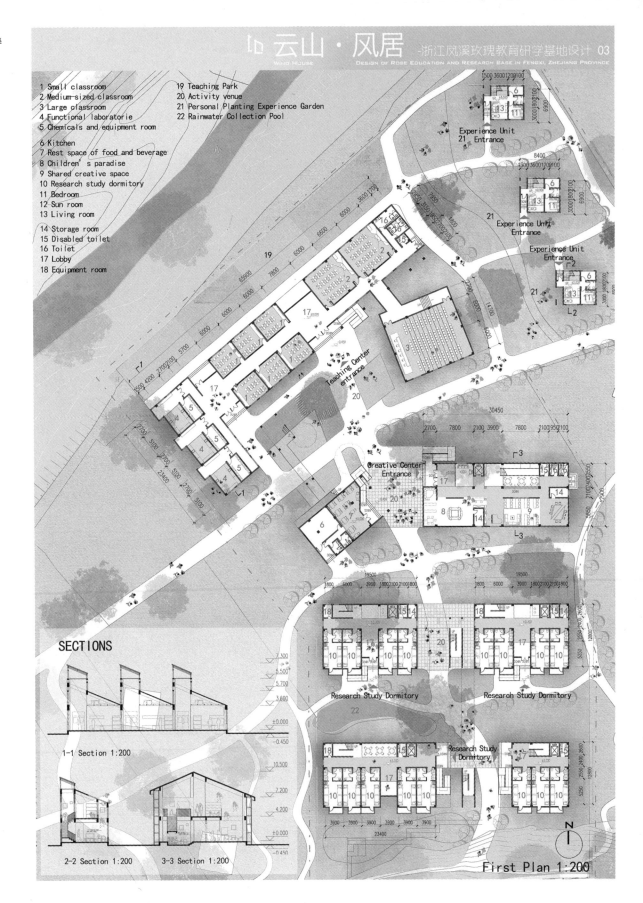

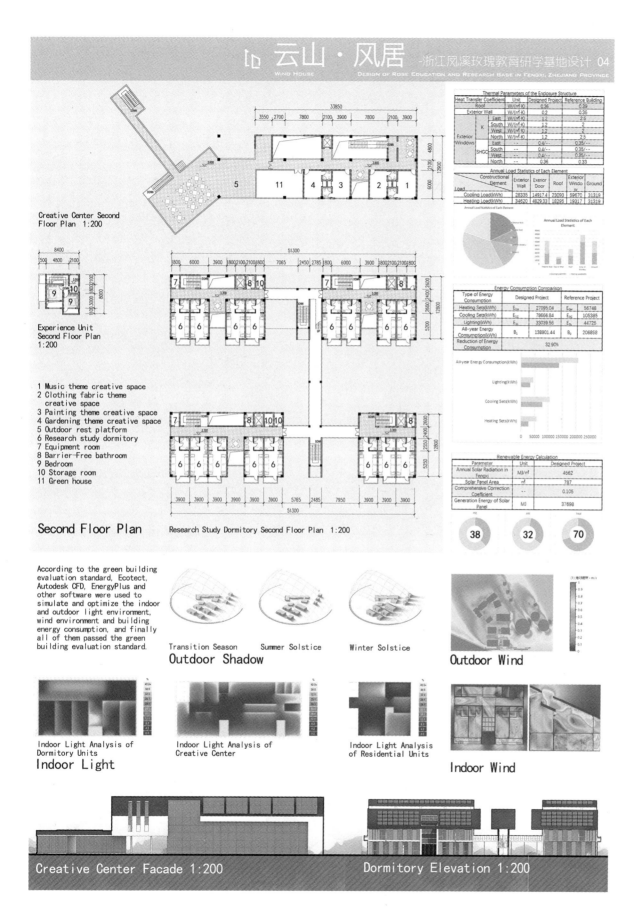

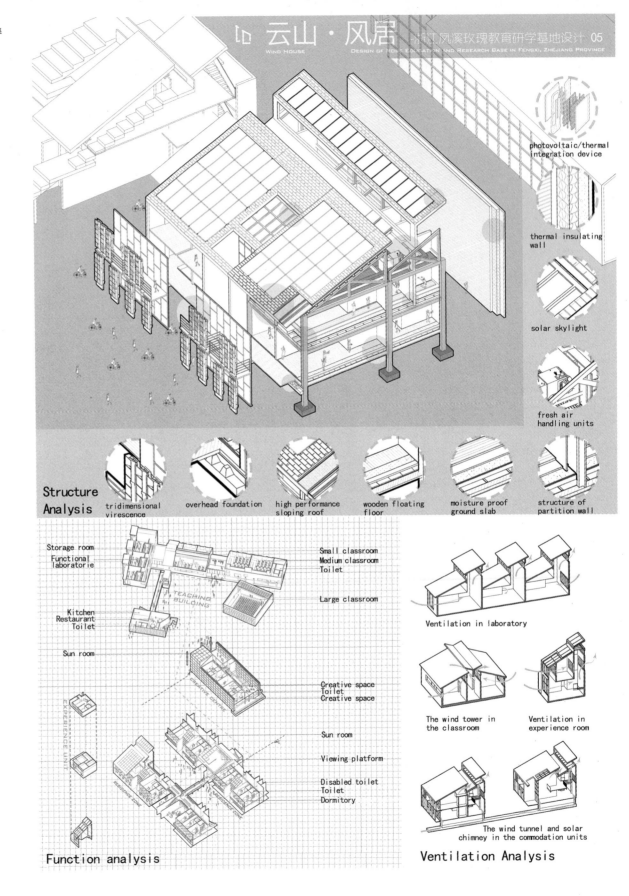

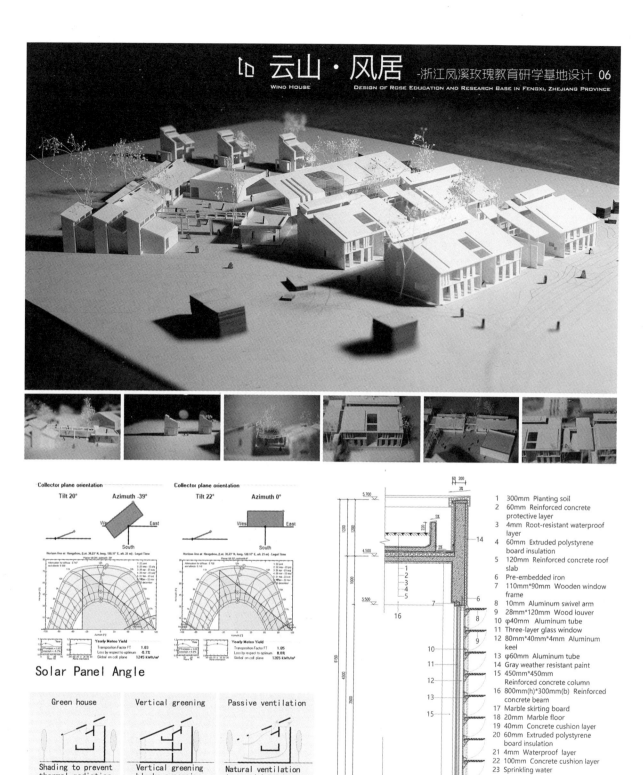

云山·风居
Wind House
-浙江凤溪玫瑰教育研学基地设计
Design of Rose Education and Research Base in Fengxi, Zhejiang Province

Solar Panel Angle

Seasonal Performance analysis

- Green house — Shading to prevent thermal radiation into rooms in summer / Heat radiation is introduced indoors in winter
- Vertical greening — Vertical greening blocks excessive sunlight in summer / Bring sunlight into the room in winter
- Passive ventilation — Natural ventilation in transition seasons / Tunnel air is introduced into rooms in summer

Shading Louver Structure in the Large Classroom 1:20

1. 300mm Planting soil
2. 60mm Reinforced concrete protective layer
3. 4mm Root-resistant waterproof layer
4. 60mm Extruded polystyrene board insulation
5. 120mm Reinforced concrete roof slab
6. Pre-embedded iron
7. 110mm*90mm Wooden window frame
8. 10mm Aluminum swivel arm
9. 28mm*120mm Wood louver
10. φ40mm Aluminum tube
11. Three-layer glass window
12. 80mm*40mm*4mm Aluminum keel
13. φ60mm Aluminum tube
14. Gray weather resistant paint
15. 450mm*450mm Reinforced concrete column
16. 800mm(h)*300mm(b) Reinforced concrete beam
17. Marble skirting board
18. 20mm Marble floor
19. 40mm Concrete cushion layer
20. 60mm Extruded polystyrene board insulation
21. 4mm Waterproof layer
22. 100mm Concrete cushion layer
23. Sprinkling water

综合奖·优秀奖
General Prize Awarded·
Honorable Mention Prize

注 册 号：6495
项目名称：游戏·园（凤溪）
　　　　　Game Garden（Fengxi）
作　　者：张思文、陈艺丹、黄晶晶、
　　　　　李彤彤
参赛单位：厦门大学
指导老师：石　峰

设计说明

本设计采用"游戏园"为主题。首先"园"体现在针对基地所处的地域环境因素和使用人群的需求规划建筑群体的布局，采用江南园林建筑的组团式布局手法，以风雨廊作为连接。"游"在图中，摆脱都市繁忙紧张的快节奏生活，在游玩中学习和体验乡村生活。建筑在太阳能被动式技术和主动式技术上相互结合的方面进行了探索，犹如一场自然环境与建筑的对抗"游戏"。在本设计中构建了一种新的建筑表皮系统，从表皮的制作、使用到维修的全过程，学生可以参与其中。同时，尝试运用智能控制系统，为节能建筑提供与时俱进的使用模式，构建健康宜居、生态发展的新农村。

Design Notes:
This design uses the theme of "Game Garden". First, the "garden" is reflected in the geographical environment factors and the needs of the use of the base, planning the layout of the building groups, using the group layout method of Jiangnan garden architecture, and using the weather corridor as a connection. We can free from the fast-paced life of the city, learn and experience rural life in the play. Buildings explored the combination of solar passive technology and active technology, like a game of natural environment and architecture. In this design, a new building skin system is constructed, from which students can participate in the whole process of making, using and repairing the skin. At the same time, we try to use intelligent control systems to provide a model for using energy-saving buildings to keep pace with the times and build a new rural area that is healthy and livable and ecologically developed.

Site Location Analysis

The project site is located in Fengxi Rose Education Base in Tonglu Country of Hangzhou City, Zhejiang Province, China.

There is a village 500 meters away from the site. Nowadays, there is a road and a river through the area.

The site is backed by mountains in the south, river in the northwest and highway in the northeast.

Concept Generalization

Divide the Area | Place Public Space | Placed Buildings | Block Dissolve | Place the Water | Corridor in Series

Climate Analysis

Hangzhou belongs to the hot summer and cold winter zone and the monsoon climate zone. In the architectural design, it is necessary to take into consideration the winter warmth and summer sunshade measures, and use the passive technology to greatly increase the building comfort. The region has sufficient solar radiation and abundant solar energy resources.

The wind speed at 1.5m in the site is less than 5m/s, and there is no obvious vortex, which meets the requirements of human comfort. The pressure difference between the building and the front reaches about 3.5pa, which is conducive to indoor ventilation.

The wind speed at 1.5m in the site is less than 5m/s, and there is no obvious vortex, which meets the requirements of human comfort.

Garden

Architectural elements in the south of Yangtze River

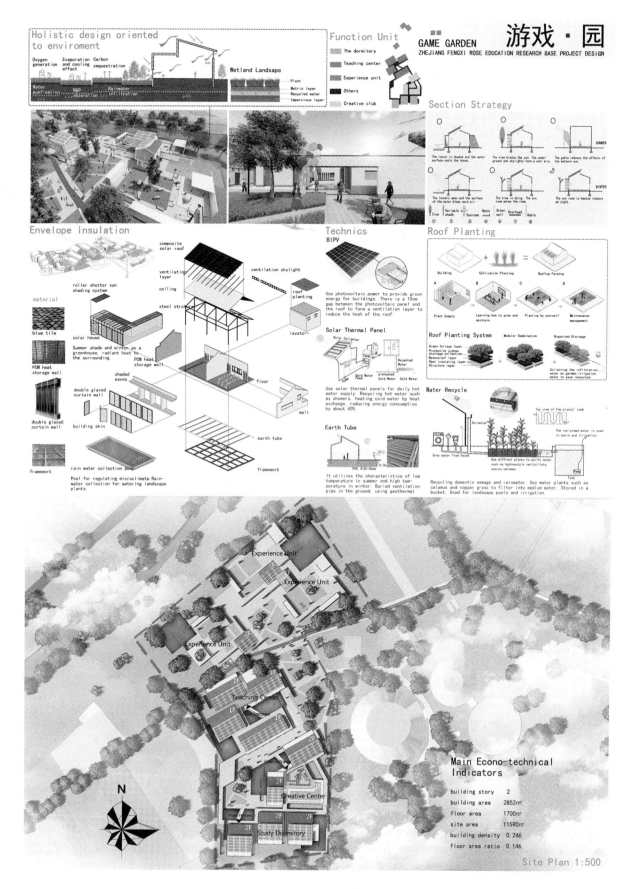

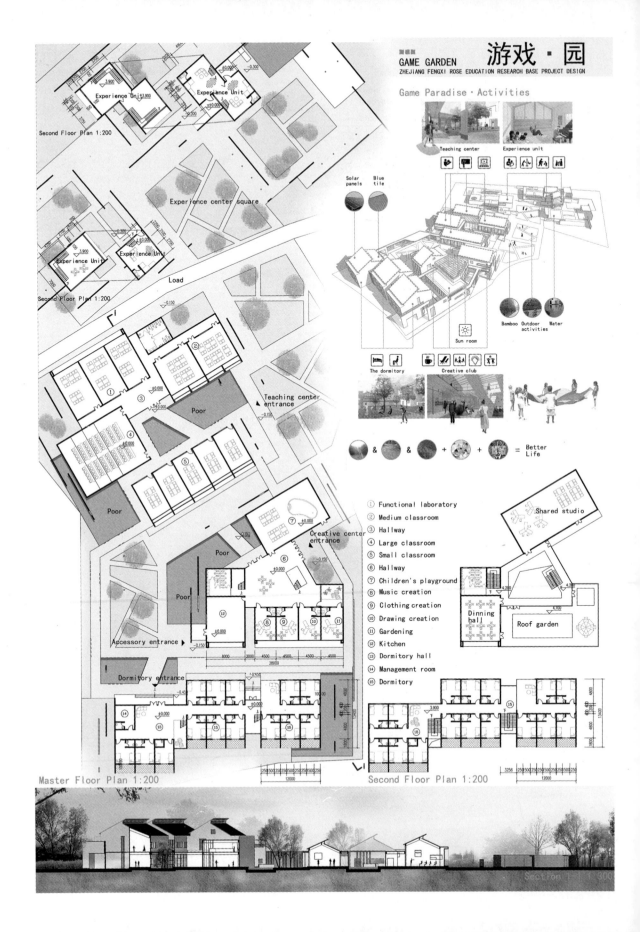

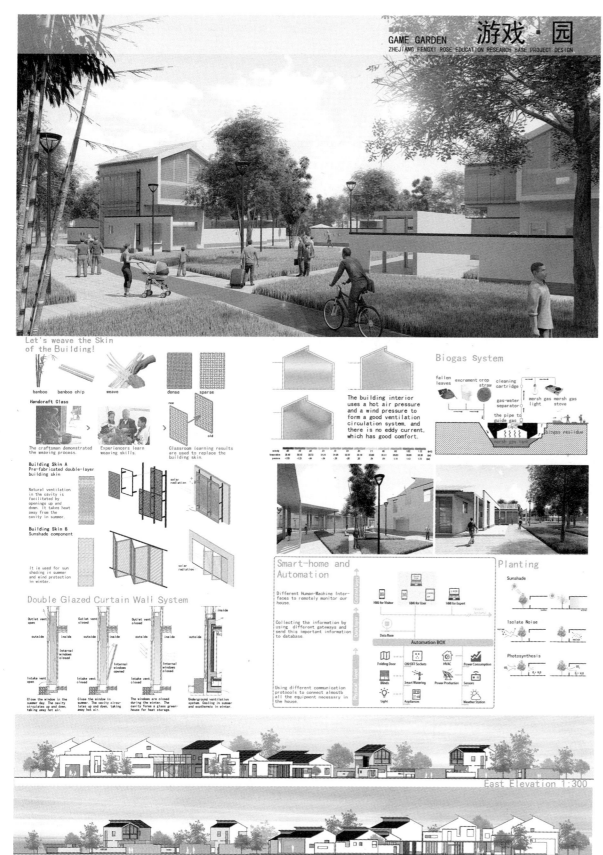

综合奖·优秀奖
General Prize Awarded · Honorable Mention Prize

注 册 号：6498
项目名称：绿色庭院研学中心（凤溪）
　　　　　Green Courtyard Research Center（Fengxi）
作　　者：郑庚锋、傅 凯、金应应、邹 倩
参赛单位：嘉兴学院
指导老师：黄 平、彭 欣、金荣科、华昕若

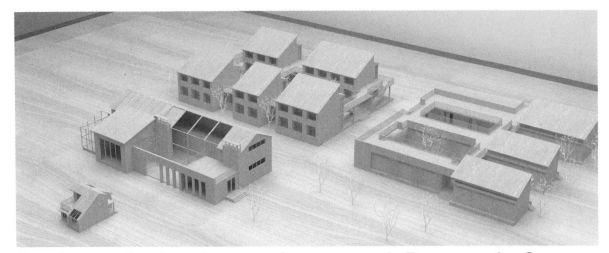

Design of the Green Courtyard Research Center

Introduction

绿色庭院研学中心设计，以江南地区的传统民区为基础进行再设计，创意中心有内庭院，中心部分利于通风，并且设有太阳能烟囱和光伏幕墙。教学中心顶部有覆土种植和架空隔热，可以上人活动游玩。研学宿舍分为五个部分，并且相互联系，有邻里的感觉，拉进人与人距离，采用改进型的特朗勃墙技术，体验单元是以家庭为主的体验乡村种植专门体验区。造型和太阳能并行，创造了节能环保的研学基地。

The green courtyard research center is redesigned on the basis of the traditional residential areas in the south of the Yangtze River. The creative center has an inner courtyard. The central part is conducive to ventilation, and has a solar chimney and a photovoltaic curtain wall. The top of the teaching center is covered with soil planting and overhead heat insulation, so that people can play. The research dormitory is divided into five parts, and they are connected with each other, have the feeling of neighborhood, pull into the distance between people and people, adopt the improved Trumbe Wall technology, and experience unit is a special experience area of family-based experience of rural planting. Modeling and solar energy are parallel, creating a research base for energy saving and environmental protection.

General site plan

Entertainment Unit

1. Studying room
2. Livingroom
3. Vestibule
4. Toilet
5. Bedroom
6. Entertainment space
7. Deck

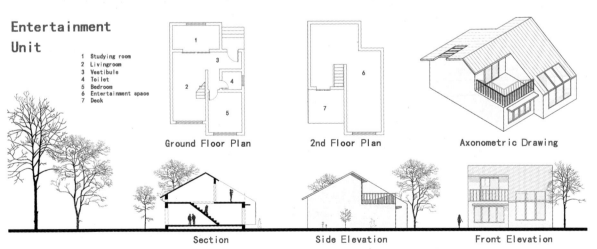

Ground Floor Plan　　2nd Floor Plan　　Axonometric Drawing

Section　　Side Elevation　　Front Elevation

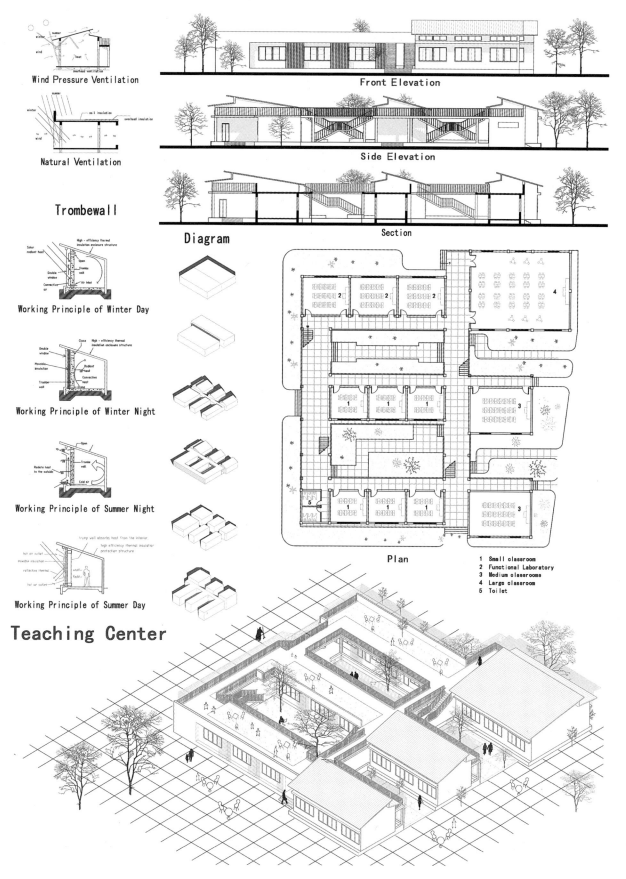

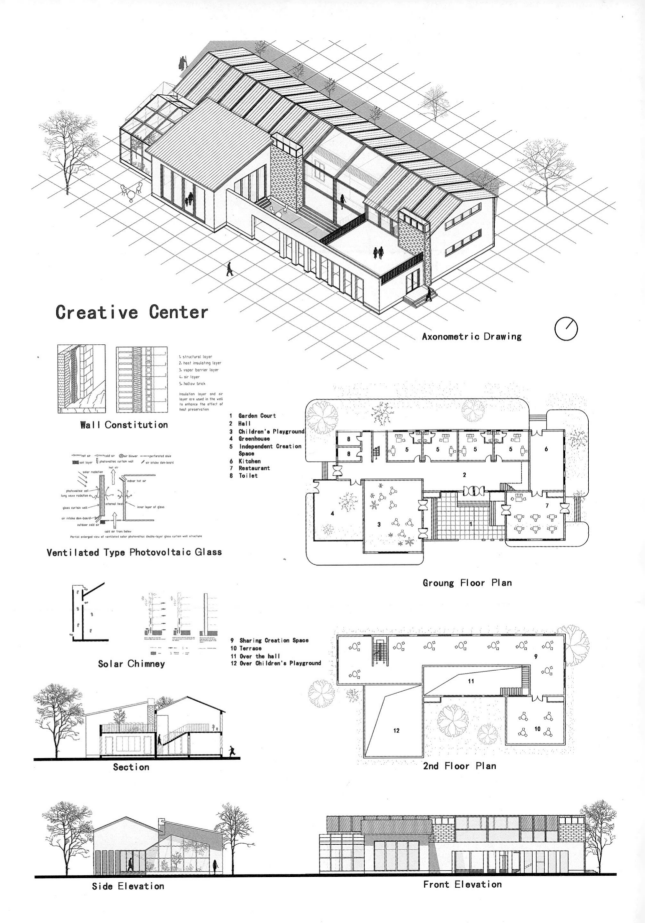

Front Elevation

Side Elevation

Study Dormitory

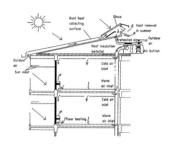

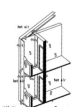

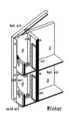

Improved Trombewall Technology

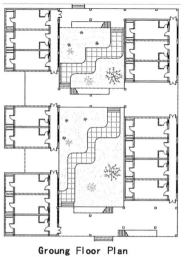
Groung Floor Plan

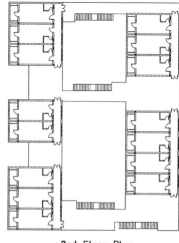
2nd Floor Plan

Section

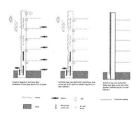

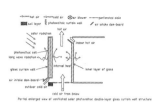

Ventilated Type Photovoltaic Glass

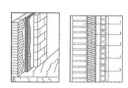

Wall Constitution

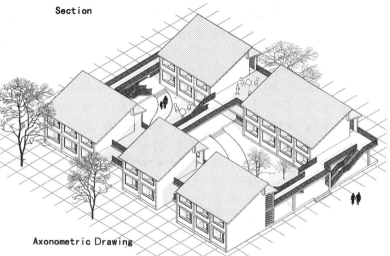
Axonometric Drawing

综合奖·优秀奖
General Prize Awarded · Honorable Mention Prize

注 册 号：6580
项目名称：呼·吸（凤溪）
　　　　　Breath（Fengxi）
作　　者：赵茂均、董　鑫
参赛单位：重庆大学
指导老师：周铁军、张海滨

呼·吸
浙江凤溪玫瑰教育研学基地

逻辑概述 Logical Overview

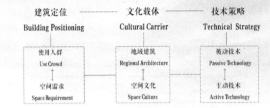

前期分析 Pre-analysis

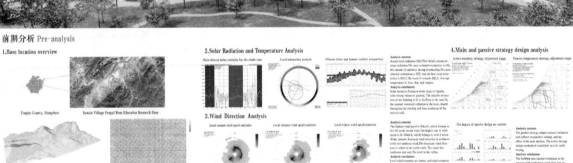

概念阐释 Conceptual Interpretation

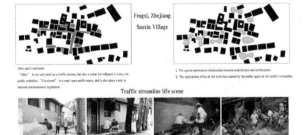

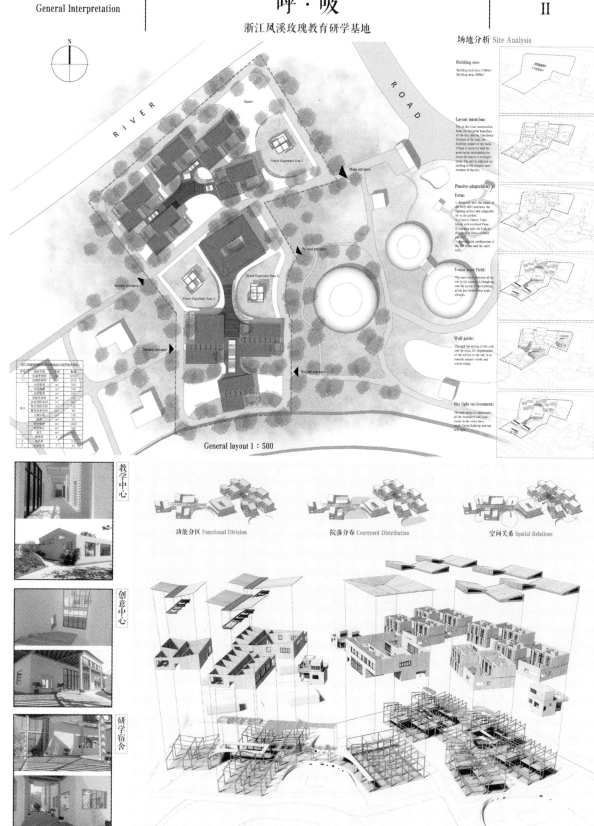

呼·吸
浙江凤溪玫瑰教育研学基地

Floor Plan | III

1. Small classroom
2. Medium-sized classroom
3. Large classroom
4. Functional laboratory
5. Bathroom
6. Shared creative space
7. Independent creative space
8. Dining break space
9. Children's playground
10. Kitchen
11. Bathroom
12. Study dormitory
13. Experience unit
14. Storage room
15. Sharing the living room

装配式体验单元 Assembled experience unit

First and Second Floor Plan 1:300

Experience unit 1

Experience unit 2

Experience unit 3

Design Strategy

呼·吸
浙江凤溪玫瑰教育研学基地

IV

呼吸策略 Respiratory Strategy

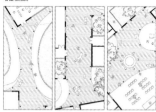

连接廊道 Connection Corridor

呼：
1. The corridor is transparent throughout, which facilitates the free passage of airflow within the site.
2. The main body of the corridor adopts a topless frame structure, and combines the planting of vine plants to bring the heat around the corridor through the transpiration of the plants.
3. The combination of the corridor and the building is covered, and the top part of the glass is matched with the adjustable visor canvas to provide an outdoor communication place for sheltering from the wind.

吸：
The vines are deciduous in winter and the visor canvas is opened to ensure sufficient solar radiation in the corridor.

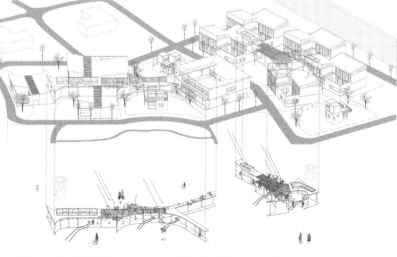

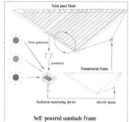
Self-powered sunshade frame

Adjustable shade sail

Connectivity and permeability

宿舍单元 Dormitory Unit

呼：
1. The roof layer is ventilated, collecting the easterly easterly wind. As the air duct shrinks at the bottom wind-extracting wall, the airflow is accelerated, the air pressure is reduced, and the wind pressure is extracted.
2. The balcony outside the perforated aluminum plate shading system (see the component analysis for the specific form), the position of the aluminum plate can be freely adjusted through the double side rail to block the summer solar radiation.

吸：
The dormitory is equipped with a south balcony, the inner wall of the balcony is a heat collecting wall, and a good heat absorbing container is formed in winter.

创意中心 Creative Center

呼：
1. The bottom floor of the creative center is partially overhead on the southeast and northwest sides, and the summer wind can pass through the building.
2. The shape of the building is "back" shaped, forming a patio. The roof is in the shape of an inverted bell mouth. When the airflow through the upper part, wind pressure can be formed.

吸：
Multiple sun rooms are embedded in the unit, such as the building entrance hall as the heat collecting chamber and the south side curtain wall sun room.

教学单元 Teaching Unit

呼：
1. The cold alley is set inside the teaching unit, which is opened in summer, and the indoor heat is taken away through the cold alley ventilation.
2. The top is equipped with a louver top window with a slope of 60° to block direct light between 10:00 and 14:00 in summer.

吸：
The classrooms are equipped with a large double-glazed window, which is good for winter heat collection (the number of solar radiation is blocked by the number in summer).

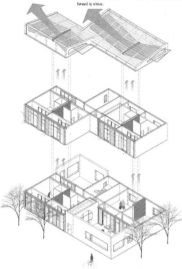

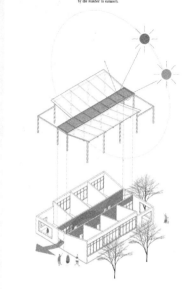

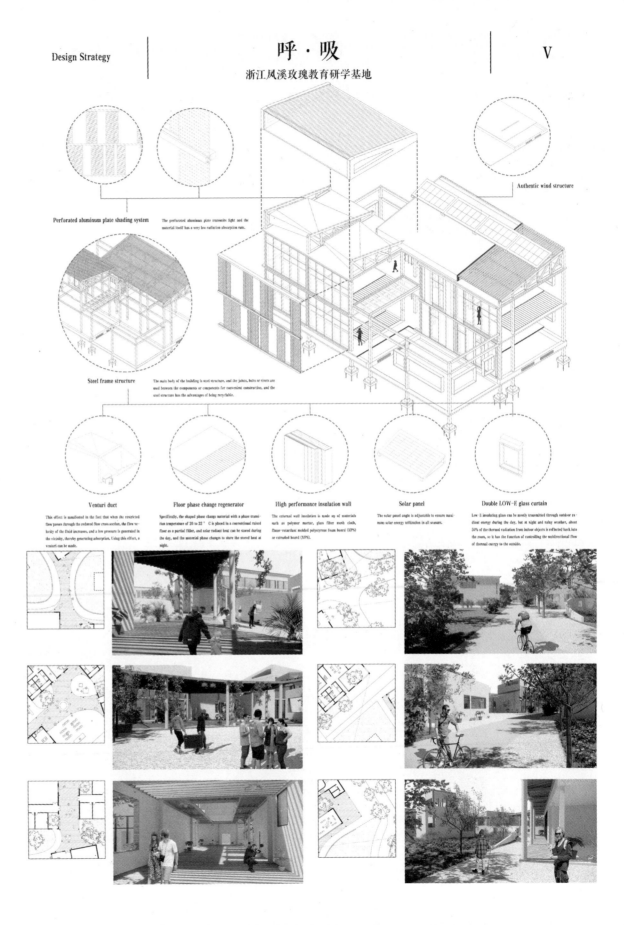

Design Strategy	呼·吸	VI
	浙江凤溪玫瑰教育研学基地	

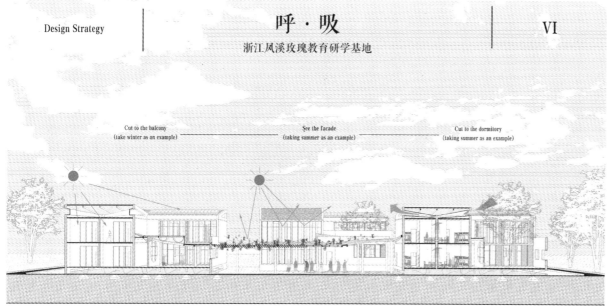

Dormitory group sectional perspective

室内光环境模拟 Indoor light environment simulation　　穿孔铝板遮阳幕 Perforated aluminum plate sunshade

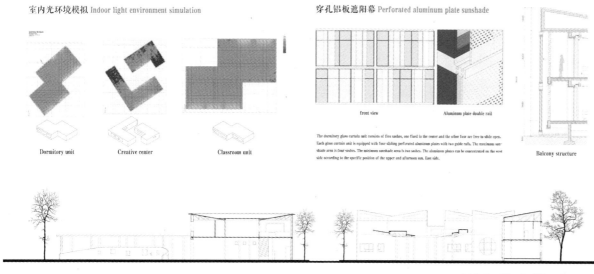

1-1 Sectional View 1∶200

South Elevation 1∶200

North Elevation 1∶200

综合奖·优秀奖
General Prize Awarded · Honorable Mention Prize

注 册 号：6583
项目名称：方·圆 光·腔（凤溪）
　　　　　Round · Square
　　　　　Green House Cavity
　　　　　(Fengxi)
作　　者：张 浩、朱强华、邓汉圆、
　　　　　和尧熙
参赛单位：华中科技大学
指导老师：徐 燊

方·圆 光·腔 01　　Round · Square Green House Cavity　　浙江桐庐凤溪玫瑰太阳能设计

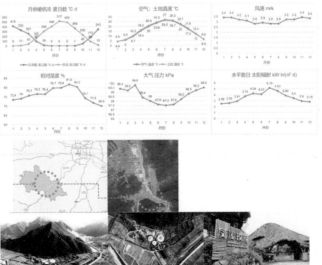

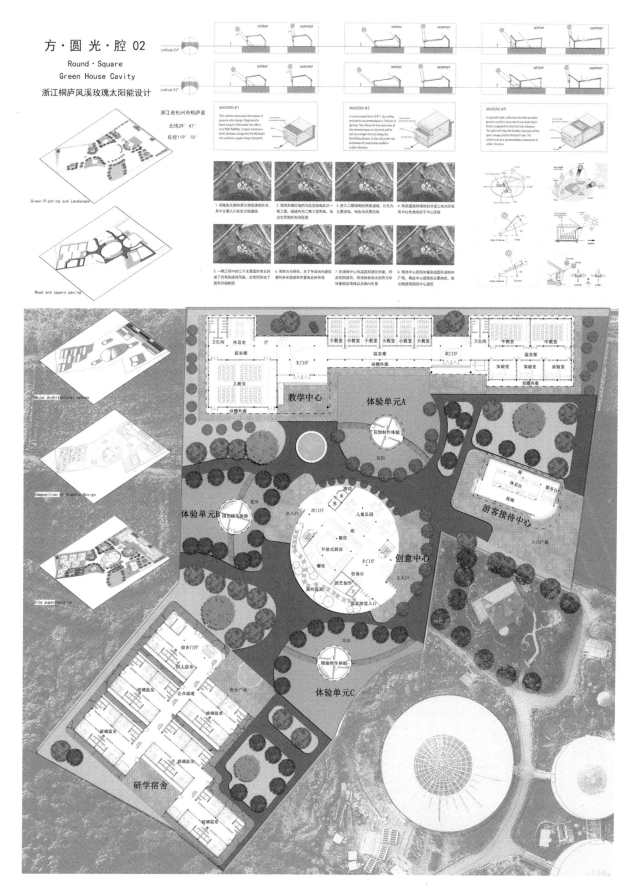

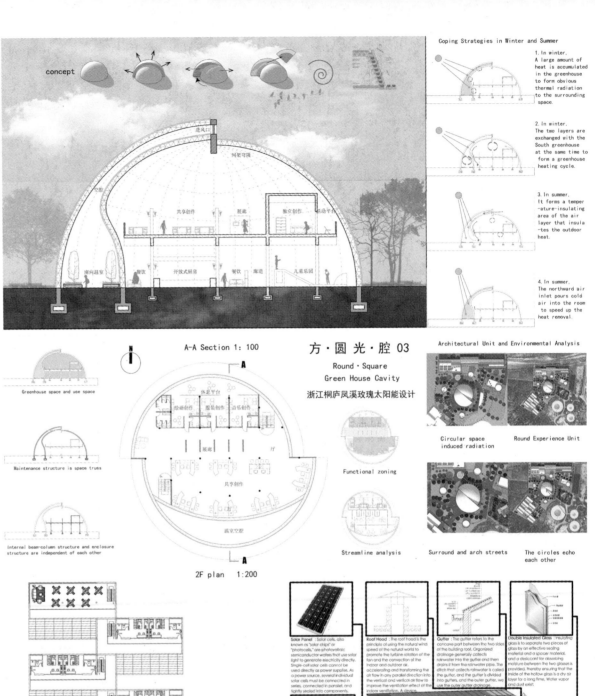

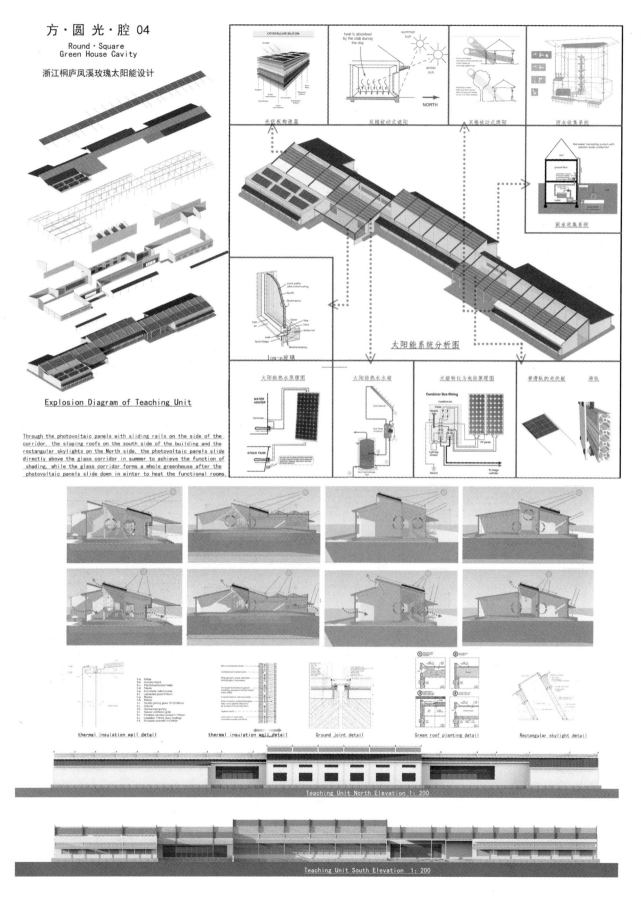

综合奖·优秀奖
General Prize Awarded · Honorable Mention Prize

注 册 号：6589
项目名称：凤源·熙聚（凤溪）
　　　　　Sunshine Base in Fengxi (Fengxi)
作　　者：赵与谦、陆思豪
参赛单位：南京工业大学
指导老师：杨亦陵、张伟郁

凤源·熙聚
Delta Cup 2019 Sunshine Base in Fengxi

Aerial view of the northwest corner of the base

Design Description

本方案以"阳光聚落"为主题，借鉴当地传统村落的形成原理，希望设计出一座能够利用太阳能创造良好的生活学习环境，融入自然的研学基地。

考虑到基地所处地区的太阳能资源有限，本设计以被动式太阳能技术为主。调研中发现当地古村落的群体布置在通风、保暖方面有着很好的特性，故将其中的挑檐、阁楼、院落、水系加以分析、改造，重组为现代背景下更开放、活跃的研学聚落，创造符合夏热冬冷地区环境的绿色、低碳的研学服务设施。

此外，光伏发电、采暖等主动式技术的合理运用，使其经济效益有效提高。

With the theme of "sunshine settlement", this program draws on the principle of local traditional villages and hopes to design a research base which can use the solar energy and incorporate into nature.
Considering the limited solar energy in the area, this design is based on passive solar technology. The survey found that the arrangement of the local villages has good characteristics in terms of ventilation and warmth, so the provocation, attic, courtyard and water system are analyzed and transformed to reorganize a more open and active research 'Settlement' and create a green, low-carbon research service facilities that meet the hot summer and cold winter environment in the modern context.
In addition, active technologies such as photovoltaic power generation and heating are also used rationally to effectively improve the economic benefits.

Climate Analysis

Best Orientation | Dry Bulb Tempture | Monthly Diurnal Average

Regional Context

Due to the wet and rainy climate in north-west Zhejiang, tradiontional residential buildings are mainly made of Bricks and Woods.

Preliminary Research on Solar Energy Design

Active Solar Energy
Active solar energy system is a forced circulation, composed of solar collector, pipeline, fan or pump, heatstorage device, etc. But the investment is large, and management is difficult.

Active Solar Energy
Passive solar energy building makes full use of solar radiation heat in winter and minimizes the heat in summer through architectural design, and achieves comfortable indoor environment.

Solar-Energy Distribuyion Drawing of Zhejiang

1300
1250
1300

Tonglu is in the 1250-1300 zone, which means passive solar energy utilization will be the main method.

Preliminary investigation · 凤

Location of the Site

Tonglu County

Location Analysis

Mountains & City & Highway | Roads & Villages | Rivers & Site

As the 'Beautiful Country' policy continuously implementing, new architectures Which designed according to the context and tradition keep built up in Tonglu.

Solar Energy Design System Representation

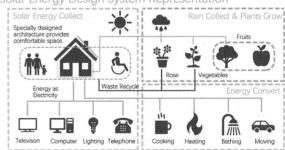

Solar Energy Collect — Specially designed architecture provides comfortable space.

Rain Collect & Plants Grow — Fruits, Rose, Vegetables

Energy as Electricity | Waste Recycle | Energy Convert

Television | Computer | Lighting | Telephone | Cooking | Heating | Bathing | Moving

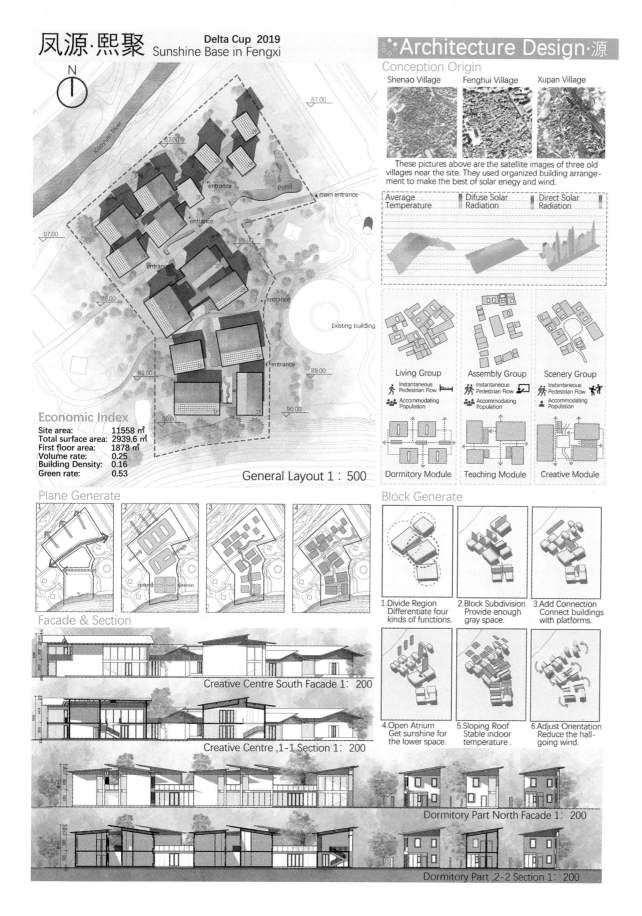

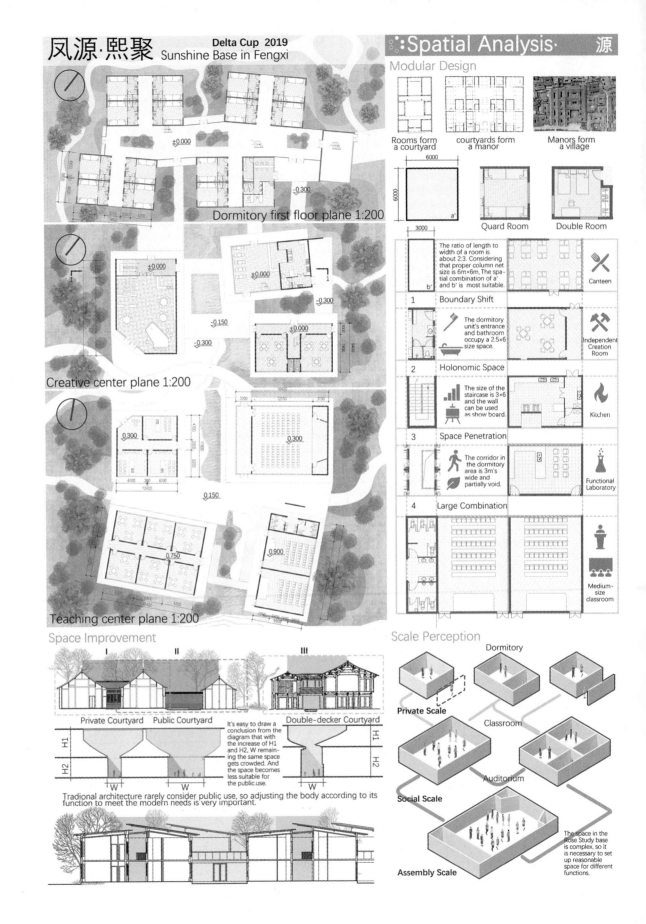

凤源·熙聚
Delta Cup 2019
Sunshine Base in Fengxi

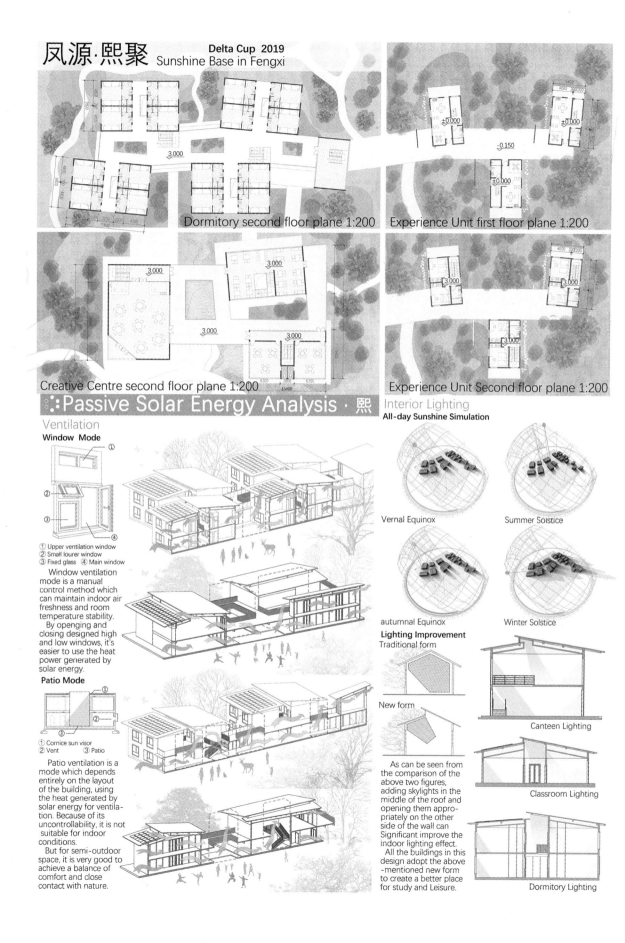

Dormitory second floor plane 1:200
Experience Unit first floor plane 1:200
Creative Centre second floor plane 1:200
Experience Unit Second floor plane 1:200

Passive Solar Energy Analysis · 熙

Interior Lighting
All-day Sunshine Simulation

Vernal Equinox | Summer Solstice
autumnal Equinox | Winter Solstice

Ventilation
Window Mode
① Upper ventilation window
② Small louver window
③ Fixed glass ④ Main window

Window ventilation mode is a manual control method which can maintain indoor air freshness and room temperature stability.
By openging and closing designed high and low windows, it's easier to use the heat power generated by solar energy.

Patio Mode
① Cornice sun visor
② Vent ③ Patio

Patio ventilation is a mode which depends entirely on the layout of the building, using the heat generated by solar energy for ventilation. Because of its uncontrollability, it is not suitable for indoor conditions.
But for semi-outdoor space, it is very good to achieve a balance of comfort and close contact with nature.

Lighting Improvement
Traditional form

New form

As can be seen from the comparison of the above two figures, adding skylights in the middle of the roof and opening them appropriately on the other side of the wall can Significant improve the indoor lighting effect.
All the buildings in this design adopt the above-mentioned new form to create a better place for study and Leisure.

Canteen Lighting
Classroom Lighting
Dormitory Lighting

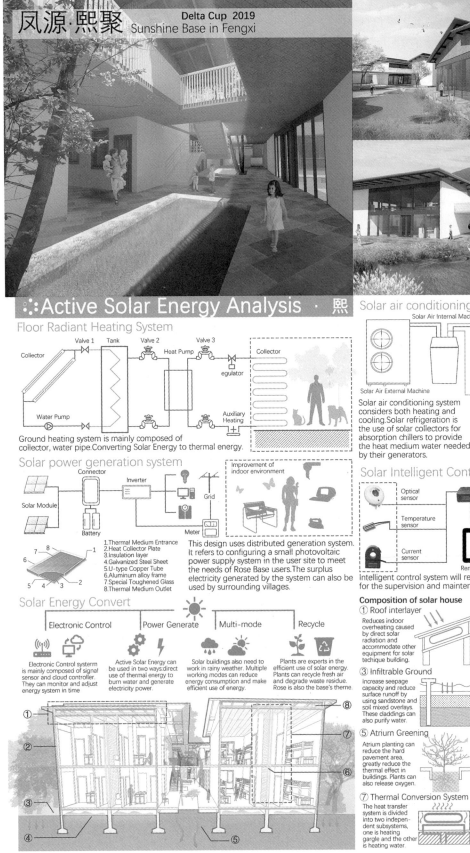

凤源·熙聚 Delta Cup 2019 Sunshine Base in Fengxi

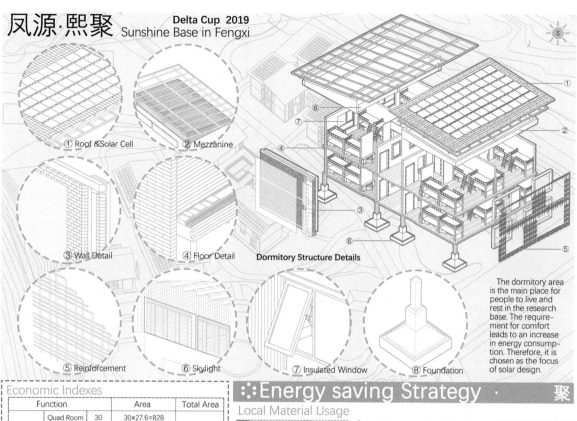

① Roof & Solar Cell
② Mezzanine
③ Wall Detail
④ Floor Detail
⑤ Reinforcement
⑥ Skylight
⑦ Insulated Window
⑧ Foundation

Dormitory Structure Details

The dormitory area is the main place for people to live and rest in the research base. The requirement for comfort leads to an increase in energy consumption. Therefore, it is chosen as the focus of solar design.

Economic Indexes

Function			Area	Total Area
Dormitory	Quad Room	30	30×27.6=828	1166.6
	Service Room	2	56.4+57.6=114	
	Sports Room	1	57.6	
	Plank Road	/	167	
Creative Center	Sharing Creation		170	707
	Independent Creation		4×35=140	
	Canteen		165	
	Children's Playground		160	
	Restroom		20	
	Kitchen		52	
Teaching Center	Small Classroom	6	6×30=180	598
	Middle Classroom	2	2×68=136	
	Big Classroom	1	192	
	Laboratory	3	3×30=90	
Experience Unit	Building		3×100=300	
Corridor		/	186	2939.6

Enclosure Structure Selection

Construction Level	Thermal Conductivity [W/m·k]
1. Exterior painting and finish	
2. External Insulation	λ=0.030
3. Bonding layer	
4. Reinforced concrete wall 200mm	λ=1.280
5. Varnishing glue 5mm	λ=0.930

The external thermal insulation structure adopted this time can basically eliminate the influence of cold and thermal bridges on various parts of the building.

Photovoltaic Calculation

Solar cell total area: 605㎡
Installation angle: 14°
Average annual sunshine intensity: 1248.3h/year
Average annual solar radiation: 4353000 MJ/m2

Σ=605*4353*0.17=447047MJ
Σ'=447047*0.28=125173kwh

According to the standard*, the average base per square meter is 40kw. Total electricity consumption is 40*2940= 117600≦Σ'. That is, the theoretical power generation is bigger than the electricity consumption.

*Refer to Chinese buildings average energy consumption standard

Energy saving Strategy · 聚

Local Material Usage

石 Pebble Stone Gravel

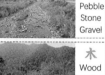

木 Wood Bamboo Reed

As Sketch — Local stones are not strong enough to build, but they are beautiful in the landscape layout, responding to nature scenery.

As Handrail — Bamboo is widely used in architecture because of its nice and tough features. The local bamboo is very suitable for making railings.

Construction Process

1. Environmental assessment
2. Foundation consolidation
3. Structural masonry

4. Roof truss assembly
5. Curtain wall and roof pavement.
6. Landscape and Equipment.

Water Conservation Measures
Improvement of aquatic ecology

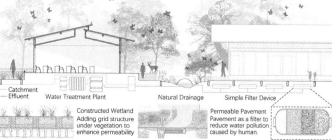

— Catchment
— Effluent

Water Treatment Plant | Natural Drainage | Simple Filter Device

Constructed Wetland Adding grid structure under vegetation to enhance permeability

Permeable Pavement Pavement as a filter to reduce water pollution caused by human.

2019 台达杯国际太阳能建筑设计竞赛获奖作品集

综合奖·优秀奖
General Prize Awarded · Honorable Mention Prize

注 册 号：5747
项目名称：星巢（兴隆）
　　　　　The Star Nest（Xinglong）
作　　者：何雨微、霍文婕、徐紫莹、
　　　　　李子儒、张颖怡、刘瑞雪
参赛单位：西安科技大学
指导老师：孙倩倩

设计采用灵活多变的六边形为元素，充分利用了有限空间，形成丰富多变的体块关系。阳光房的设置与居住单元结合紧密，在平立面上形成虚实结合的视觉效果，把组团象征为点并按照星系排列，每一个组团都隐喻一颗北斗七星，夜幕降临，灯光与星光交汇融合，在总体效果上呈现了不一样的星空。

Flexible hexagons are used as elements of design, making full use of the limited space, and which forms a rich and varied volume relationship. The setting of the sunshine room is closely combined with the living units, forming a visual effect of combining the real and the virtual on the flat facade. The cluster symbol is the point and arranged in accordance with the galaxy. Each cluster is one of the Big Dipper. As night falls, the lights merge with the stars, presenting a different starry sky in the overall effect.

BIRD'S EYE VIEW

BACKGROUND ANAYSIS
HEBEI IMAGE ANALYSIS

LOCATION ANALYSIS

Base is located in chengde city, hebei province six river ZhenAn camps xinglong village night park, north latitude 40°41', longitude 117°25'.

CLIMATE ANALYSIS

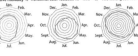

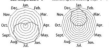

ANALYSIS OF THE CURRENT SITUATION OF THE BASE

location of the park | Road of the park | Existing building | The star bridge

THE METEOROLOGICAL

Annual average temperature between 6.5 10.3 °C, annual average rainfall of 688.9 millimeters. Xinglong county is mountainous, the vertical temperature change is obvious. Winter is northwest monsoon, summer is southeast monsoon.

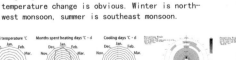

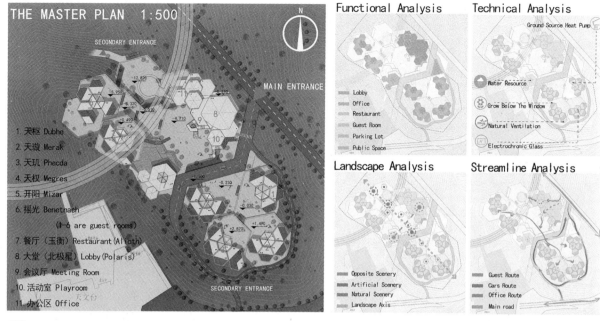

The concept of a galaxy...

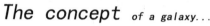

In a star-gazing park, the hotel's design was inspired by the big dipper, which introduced the galaxy. Taking each sunshine room node as a star, the general layout is set according to the position of the big dipper, and the whole park is connected by this concept.

THE GENERATION PROCESS OF THE BLOCK

星巢
THE STAR NEST

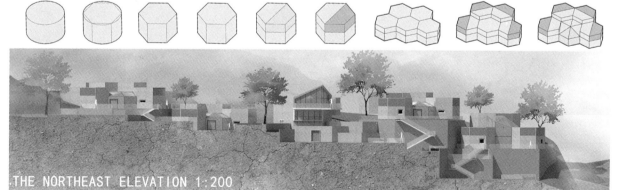

THE NORTHEAST ELEVATION 1:200

THE SOUTHWEST ELEVATION 1:200

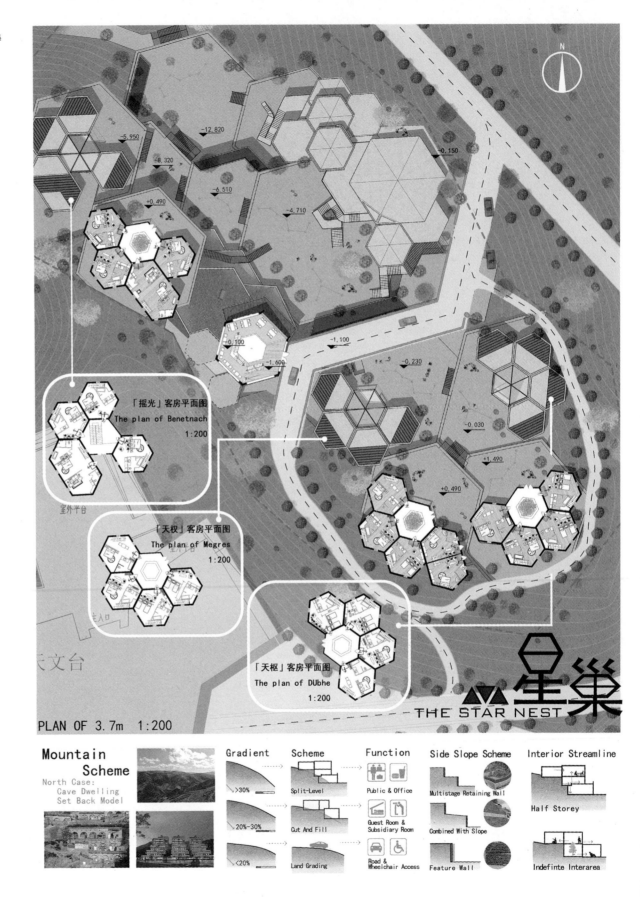

SECTIONAL PERSPECTIVE OF THE RESTAURANT

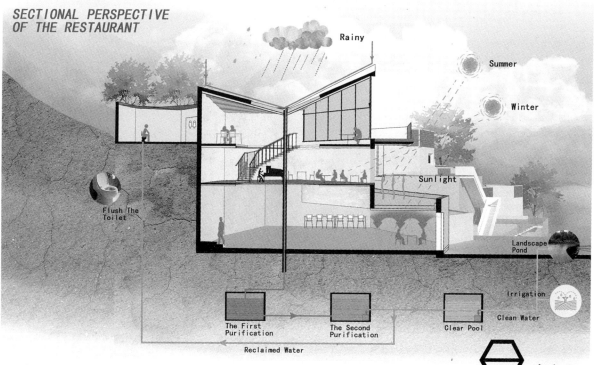

GROUND SOURCE HEAT PUMP

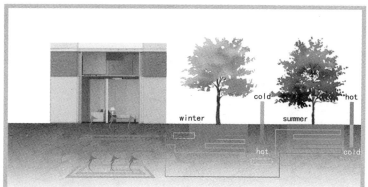

THE STAR NEST 星巢

Basement plan of restaurant 1:200

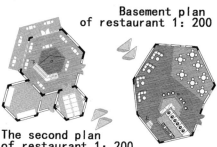
The second plan of restaurant 1:200

ELECTROCHROMIC GLASS INSTALLATION METHOD

THE RESIDENTIAL UNIT's SECTION 1:200

COLOR-CHANGE PRINCIPLE

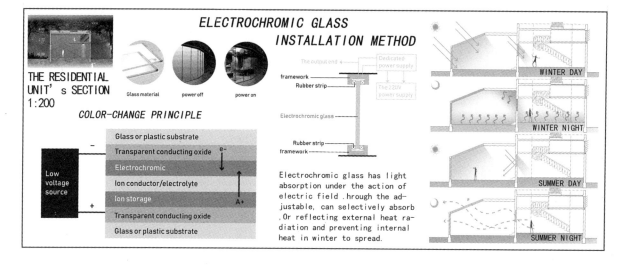

Electrochromic glass has light absorption under the action of electric field. hrough the adjustable, can selectively absorb, Or reflecting external heat radiation and preventing internal heat in winter to spread.

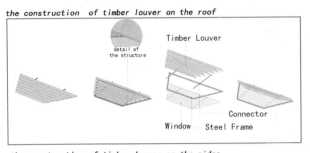
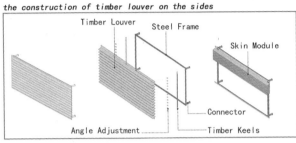
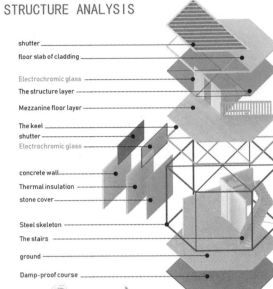
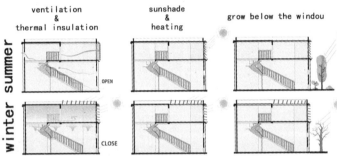

Perspective of the lobby

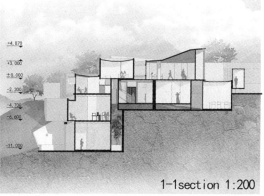
1-1 section 1:200

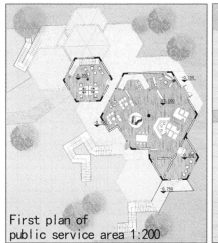
First plan of public service area 1:200

Basement plan of public service area 1:200

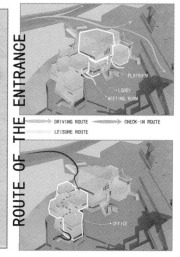
ROUTE OF THE ENTRANCE

经济分析 Economic Analysis			经济技术指标 Technical-economical Index	
功能 Function		总面积 (m²)	占地面积 Cover area	4940.49 m²
客房部分 Guest room part	标准间 Standard room	1988.67	建筑基底面积 Building area	2161.09 m²
	大床房 Double room			
	套间 Suite			
	阳光房 Sunlight room			
	观星平台 Viewing platform			
	客房服务单元 Room services			
公共部分 Common part	酒店大堂 Hotel lobby	451.7	总建筑面积 Overall floorage	3159.72 m²
	大堂服务配套 Lobby service room			
	会议室活动室 Meeting room		容积率 Volume fraction	0.6
餐饮部分 Food and beverage part	厨房 Kitchen	463.25		
	餐厅 Restaurant		建筑密度 Building density	43.70%
	酒吧 Bar			
行政部分 Administrative part	办公室 Office room	181.4	绿化率 Greening rate	15.30%
后勤部分 Logistics part	后勤服务用房 Logistics service room	101.7		

星巢
THE STAR NEST

Perspective of the restaurant

Perspective of the guest room unit

综合奖·优秀奖
General Prize Awarded · Honorable Mention Prize

注 册 号：5790
项目名称：光·星院（兴隆）
　　　　　Sunlight · Watch the Stars of the Garden（Xinglong）
作　　者：胡庆强、刘思遥
参赛单位：重庆大学
指导老师：周铁军、张海滨

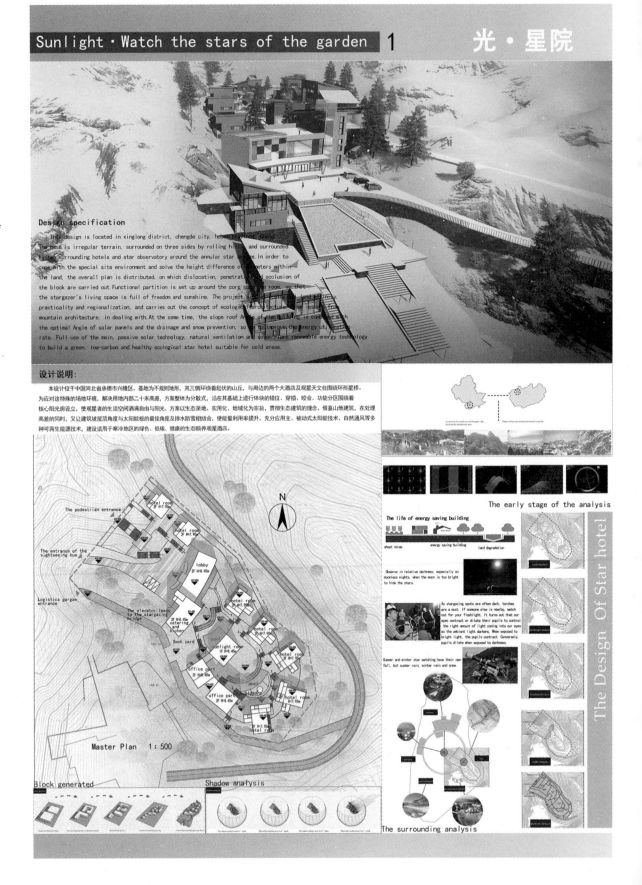

Sunlight · Watch the stars of the garden 2　光·星院

The Design Of Star hotel

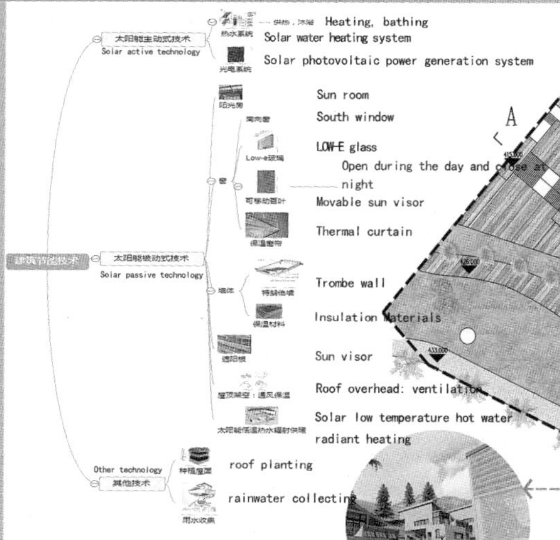

Architectural technical mind mapping

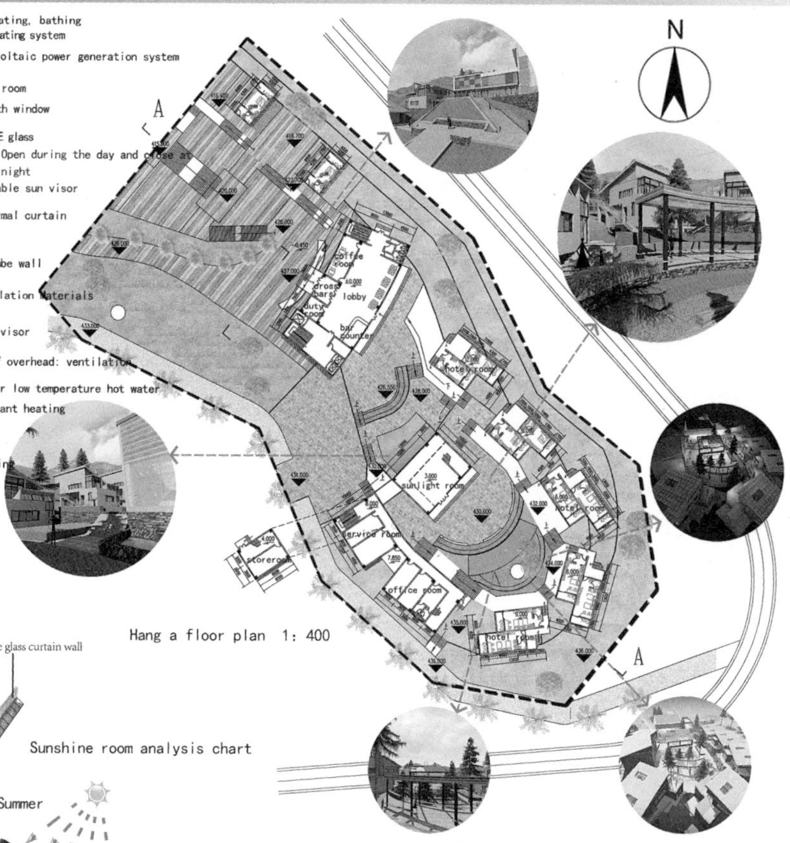

- Solar active technology
  - Solar water heating system — Heating, bathing
  - Solar photovoltaic power generation system
- Solar passive technology
  - Sun room
  - South window
  - Low-E glass
  - Open during the day and close at night
  - Movable sun visor
  - Thermal curtain
  - Trombe wall
  - Insulation materials
  - Sun visor
  - Roof overhead: ventilation
  - Solar low temperature hot water radiant heating
- Other technology
  - roof planting
  - rainwater collecting

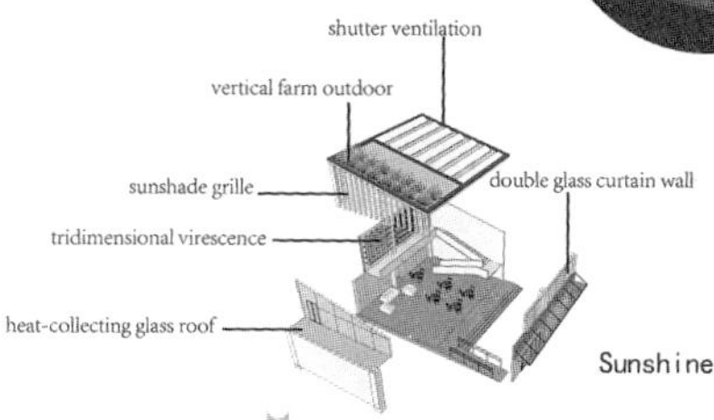

Sunshine room analysis chart

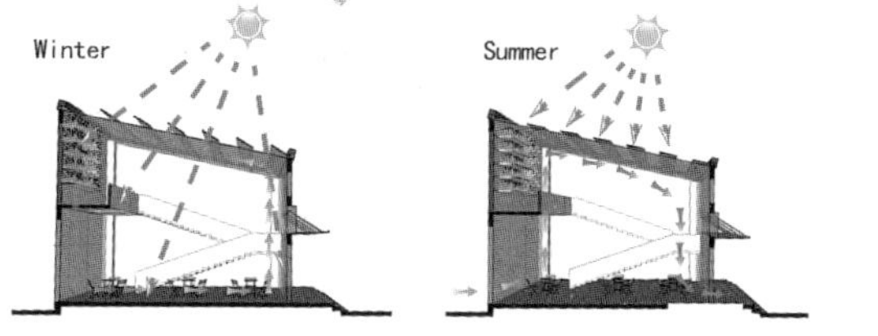

- shutter ventilation
- vertical farm outdoor
- sunshade grille
- double glass curtain wall
- tridimensional virescence
- heat-collecting glass roof

Hang a floor plan 1:400

Plan of the first floor 1:400

Winter
The sun visor is opened during the day, and the sunlight enters to heat the indoor temperature;

The sun visor is closed at night, and the heat is radiated to the room through the heat storage floor and the wall;

Summer
The sun visor is turned off during the day to reduce heat ingress and pass through the plants on the wall to reduce the temperature of entering the indoor air;

Open the sun visor at night and use the cold air to remove excess heat from the room;

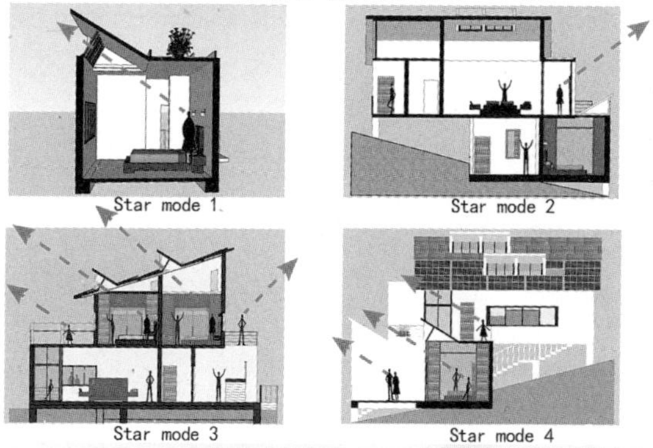

Star mode 1　Star mode 2　Star mode 3　Star mode 4

Northeast facade 1:250

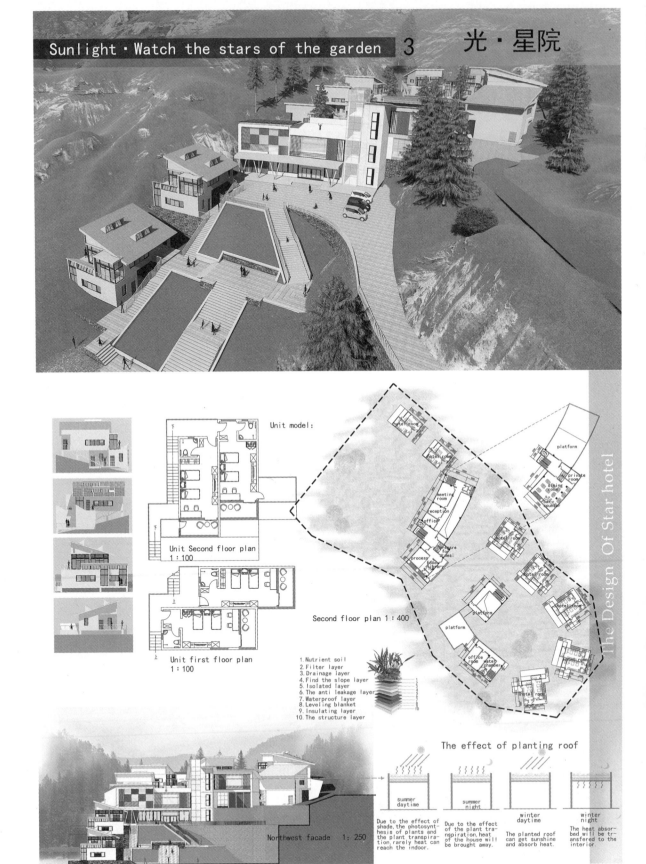

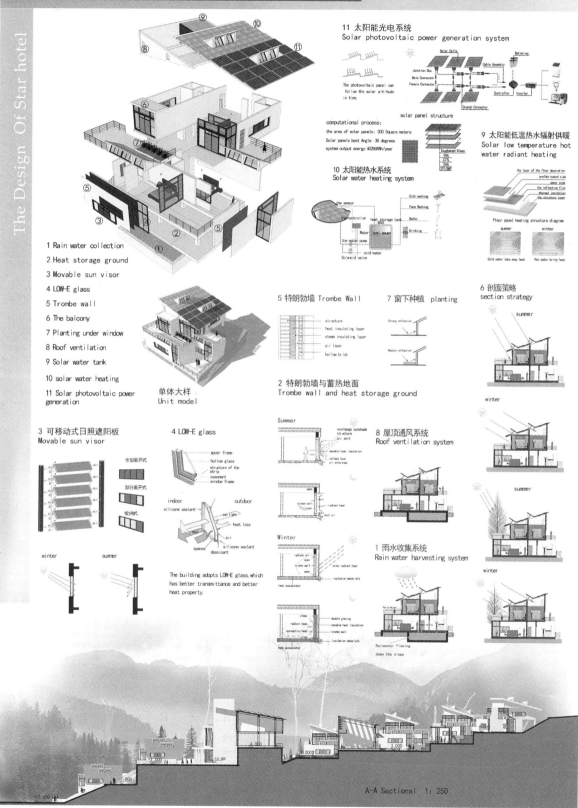

综合奖·优秀奖
General Prize Awarded · Honorable Mention Prize

注 册 号：5938
项目名称：星·桥·栈（兴隆）
　　　　　Star · Bridge · Stack
　　　　　（Xinglong）
作　　者：王杰汇、马振雷、姜子信、
　　　　　巫恋恋、杨心悦
参赛单位：天津大学
指导老师：郭娟利

星·桥·栈　STAR · BRIDGE · HOTEL　01

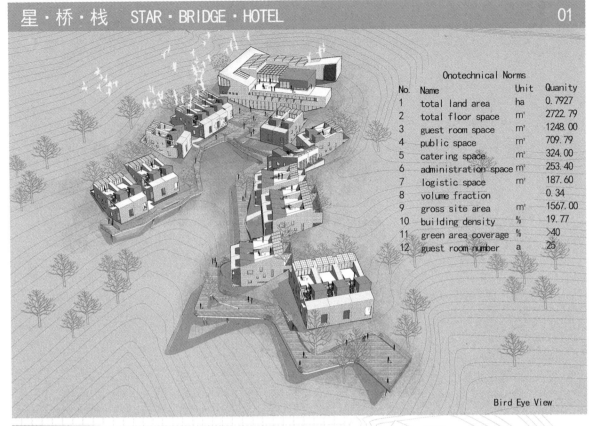

No.	Name	Unit	Quanity
1	total land area	ha	0.7927
2	total floor space	m²	2722.79
3	guest room space	m²	1248.00
4	public space	m²	709.79
5	catering space	m²	324.00
6	administration space	m²	253.40
7	logistic space	m²	187.60
8	volume fraction		0.34
9	gross site area	m²	1567.00
10	building density	%	19.77
11	green area coverage	%	>40
12	guest room number	a	25

Bird Eye View

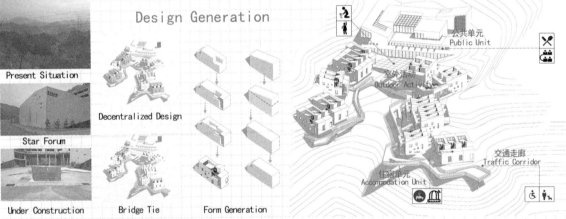

Site Plan

设计说明：
　　设计原型的来源基于公园本身的山地特点，采用"化整为零"的设计和建造方式，采用"小规模、组团式、生态化、微田园、阶梯式"的建造模式，实现地形和单元的有机结合，"见缝插绿"，足不出户便可尽享微田园风光并形成不同的田园单元模式。

Design Description:
The source of the design prototype is based on the mountain characteristics of the park itself. It adopts the design and construction method of "zeroing out" and adopts the "small scale, group, ecological, micro pastoral, stepped" construction mode to realize the terrain and unit. Organic combination, "seeing the green in the seams", you can enjoy the micro-idyllic scenery without leaving the house, and form different rural unit modes.

星·桥·栈 STAR·BRIDGE·HOTEL

Public Unit Plan

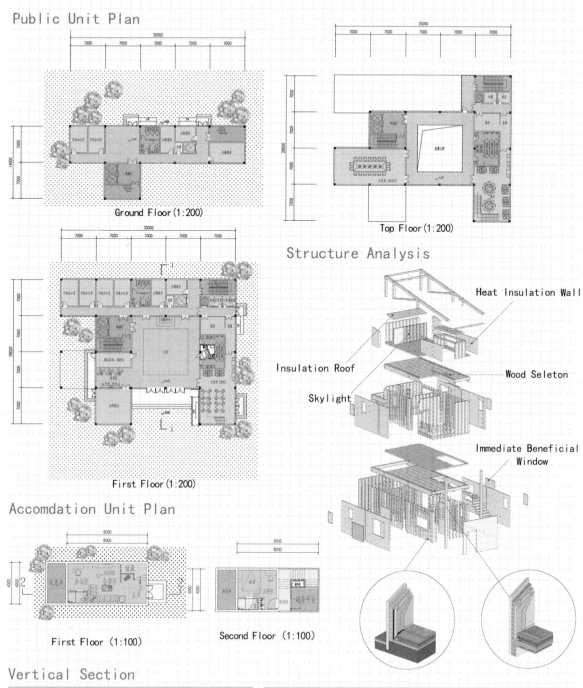

Ground Floor (1:200)

Top Floor (1:200)

Structure Analysis

First Floor (1:200)

- Heat Insulation Wall
- Insulation Roof
- Wood Seleton
- Skylight
- Immediate Beneficial Window

Accomdation Unit Plan

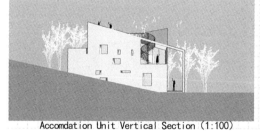

First Floor (1:100)

Second Floor (1:100)

Vertical Section

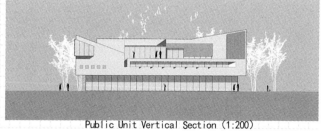

Accomdation Unit Vertical Section (1:100)

Public Unit Vertical Section (1:200)

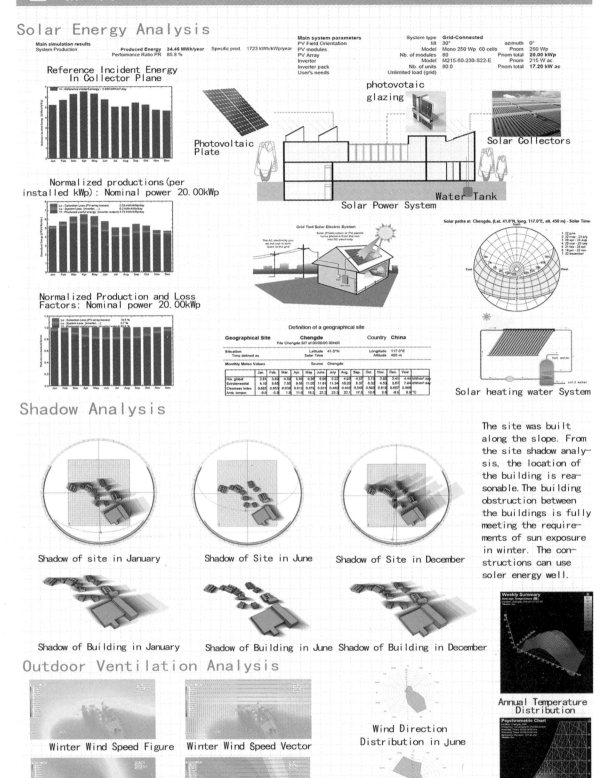

星·桥·栈 STAR·BRIDGE·HOTEL

Daylighting Analysis

Public Unit Lighting Effects

Top Floor illumination

Skylight

First Floor illumination

Bedroom

Top Floor illumination

Side-window

First Floor illumination

Bedroom

The public unit adopts a flat skylight window to improve the natural illuminance of the room, and the three-story winter solstice daylighting meets the illumination requirements

Therml Analysis

Ground-source heat pump

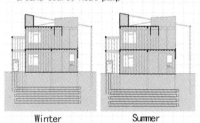
Winter   Summer

Ground-source heat pump is a heating and cooling air-conditioning system that uses solar energy resources stored in soil as a source of cold and heat for energy conversion.

Passive ventilation strategy

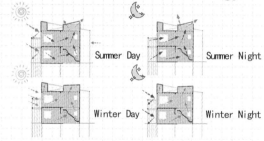
Summer Day   Summer Night
Winter Day   Winter Night

Section

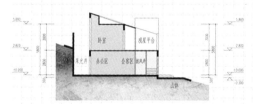
2-2 Section (1:200)

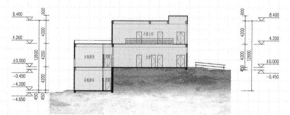
1-1 Section (1:200)

综合奖 · 优秀奖
General Prize Awarded · Honorable Mention Prize

注 册 号：5942
项目名称：星光暗夜·山居秸院（兴隆）
Starlight Dark Night ·
Hillhouse Straw Courtyard
(Xinglong)
作　　者：曹锦波、蒋群歆、徐　敏、
　　　　　肖　歆
参赛单位：广州大学、华南理工大学
指导老师：李　丽、周孝青

星光暗夜·山居秸院 1
Starlight Dark Night · Hillhouse Straw Courtyard

Technical-Econmic Index	
Total area	2719.51 m²
Site area	7940.57 m²
Building foot print	700.82 m²
Floor area ratio	24.69%
Number of guest rooms	26
Green rate	46%

01 BACKGROUN ANALYSIS

设计说明 Design Description

针对基地气候、地形地貌，规划上采用组团式结合单元式的布局方式，并形成半围合院落，改善场地内微气候；在建筑设计上，设计四种客房适应地形，降低土方量。被动设计上，南向设置较大面积开窗，冬季可以获得太阳热能。主动设计上，采用屋面太阳能集热系统，进行地板采暖、加热新风等。产业化上，采用装配式框架大板结构和模数化建造，采用当地秸秆作为围护结构的主要材料来源，钢材为承重结构，做到选材上因地制宜。

Under consideration of climatic characteristics and landform in the building site, Group layout combined with unit layout is adopted, forming a semi-enclosed compound, which improves the site microclimate. In order to adapt the landform and reduce earthwork, four kinds of units of guest rooms are designed. The large window on the south wall can bring light and warmth into the heart of the house in winter. As a solar collector, the rooftop is used to collect solar energy for wintertime space or floor heating and indoor ventilation. In industrialization, the assembly frame board structure and modular construction are adopted, the local straw is used as the main material source of the enclosure structure, and the steel is used as the load-bearing structure, so that the material selection can be adapted to local conditions.

● Site Loction Analysis

The project site is located in the Night Park of Anying Village, Liudaohe Town, Xinglong County, Chengde City, Hebei Province (under planning). It has a latitude of 40 41', an east longitude of 117 25', 115 kilometers from the center of Beijing and 30 kilometers from Xinglong County Town.

● Actuality Analysis

■ Analysis of Peripheral Traffic　■ Analysis of Peripheral Resources　■ Wind Direction Analysis of Site　■ Slope analysis of site

■ Scene photos　The project land has dense vegetation and beautiful surface landscape. Also, the area is rich in stones.

● Climate Analysis

· The average annual temperature ranges from 6.5 to 10.3 degrees Celsius, January is the coldest month of the year, the average minimum temperature is -9.9 degrees Celsius, July is the hottest month of the year, the average temperature is 23.3 degrees Celsius.

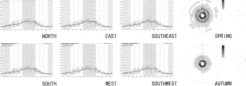

NORTH　EAST　SOUTHEAST　　SPRING　SUMMER
SOUTH　WEST　SOUTHWEST　　AUTUMN　WINTER

From the different directions of solar radiation shown above, the radiation from south, southwest and southeast is higher, and fluctuates greatly in autumn and winter.

In terms of wind direction and frequency, the southeast wind dominates in summer and the northeast wind dominates in winter.

· Shadow analysis

· Summer solstice (morning)　· Summer solstice (afternoon)　· Winter solstice (morning)　· Winter solstice (afternoon)　· Evaluation

bad / good

According to the analysis of the winter solstice shadow in the site, the northern part of the site is seriously sheltered and the sunshine hours are less than 2 hours. Therefore, the buildings should be avoided in the northwestern part of the site.

● Architectural Culture Analysis

· Hebei traditional architectural culture -- Stone Kiln

Design reference:
1. Using semi-burial method to reduce the contact area with air and achieve thermal insulation.
2. Easy to combine with topography and make full use of topography

● Local Material Analysis

· Straw Material Situation

This year, Hebei has been criticized for the haze caused by straw burning all the year round. Therefore, this project considers combining the characteristics of straw to make corresponding attempts. On the one hand, using straw material is local material, economical and practical, on the other hand, straw material is a renewable green building material, which conforms to the concept of green and sustainable development in rural areas.

● Straw Material Application

■ Straw brick/board　■ Straw wall　■ Straw House　■ Straw Floor　■ Straw Column

The application of straw material in the field of construction has been mature, and it can be used to make various parts of the building. The application of straw material conforms to the green and sustainable development goals of the countryside.

星光暗夜·山居秸院 2
Starlight Dark Night Hillhouse Straw Courtyard

02 DESIGN STRATEGY

Design Thinking

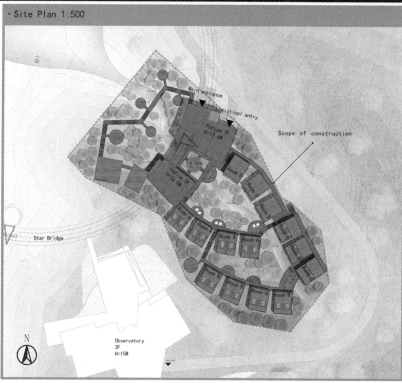

· Site Plan 1:500

Planning Strategy

· Concept

1. Creating Waterbelt Landscape, Responding to Galaxy Culture
2. Constructing star atmosphere by lighting arrangement, using scattered light sources as far as possible, avoiding pollution of large light sources, facilitating star-watching activities, and creating a brilliant artistic conception of the Milky Way stars, making people feel like walking in the middle of the Milky Way.

· Spatial Pattern Generation Process

1. Place guest rooms according to terrain and wind direction
2. Creating a Waterscape Flowing along the Terrain
3. Setting up skirt houses beside Xingqiao
4. Building a Platform in the Galaxy
5. Joining the Road System and Final Achievements

· Spatial Scale Exploration

Small scale — Star-watching private courtyard space
Medium scale — Platform shared space
Large scale — Hotel Central Courtyard Space
Space with abundant layers

· Valley Wind Analysis

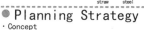
1. The Valley is open to the northwest, and the northwest wind prevails in winter.

3. Cold radiation acts on the air at night in summer, mainly moving towards low displacement along the valley slope.

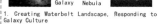
2. The southeast direction of the base is blocked by mountains, which makes it difficult for the summer monsoon to enter.

4. During daytime in summer, the air is affected by solar thermal radiation and mainly moves along the valley slope towards high displacement.

· Shadow Analysis

Summer solstice morning / Summer solstice afternoon
Winter solstice morning / Winter solstice afternoon

Building layout avoids the situation of mutual sunshine. In winter solstice venue, the northern rooms are less sheltered and can get enough sunshine. The Southern rooms are more sheltered, but the winter solstice day can also keep more than 2 hours of sunshine.

· Exploration of Planning Rationality

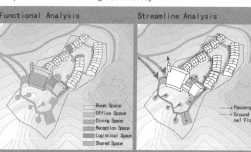 

Functional Analysis | Streamline Analysis | Microenvironment Analysis

Room Space / Office Space / Dining Space / Reception Space / Logistical Space / Shared Space

1. Public function space and private space function are separated from each other, but they form a circular relationship and are closely related and convenient.
2. The overall external streamline planning is reasonable and passenger streamline and logistics streamline do not interfere with each other.
3. Considering the topography and climate, three core landscape points are created in the planning layout, and three micro-environment spaces affecting the architecture are formed at the same time.

星光暗夜·山居秸院 3
Starlight Dark Night Hillhouse Straw Courtyard

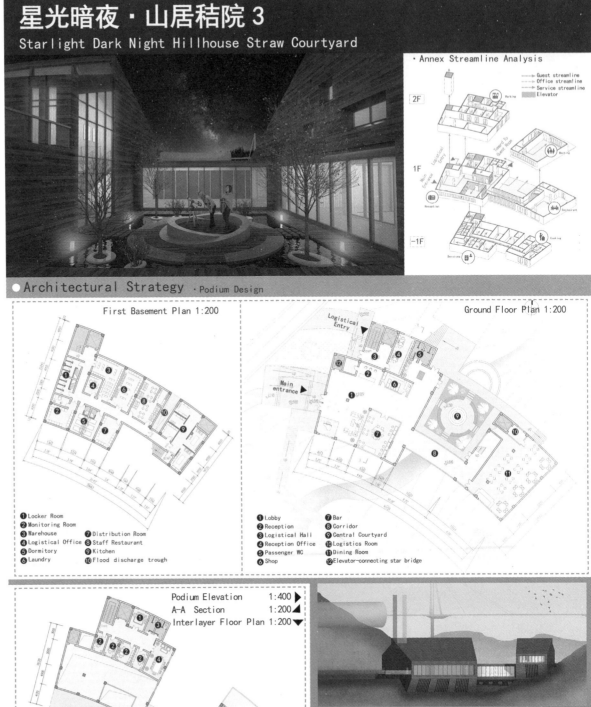

• Annex Streamline Analysis

● Architectural Strategy · Podium Design

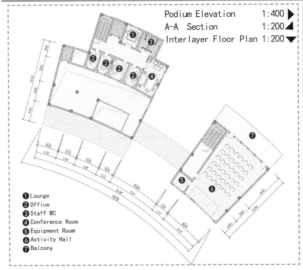

First Basement Plan 1:200

❶ Locker Room
❷ Monitoring Room
❸ Warehouse
❹ Logistical Office
❺ Dormitory
❻ Laundry
❼ Distribution Room
❽ Staff Restaurant
❾ Kitchen
❿ Flood discharge trough

Ground Floor Plan 1:200

❶ Lobby
❷ Reception
❸ Logistical Hall
❹ Reception Office
❺ Passenger WC
❻ Shop
❼ Bar
❽ Corridor
❾ Central Courtyard
❿ Logistics Room
⓫ Dining Room
⓬ Elevator-connecting star bridge

Podium Elevation 1:400
A-A Section 1:200
Interlayer Floor Plan 1:200

❶ Lounge
❷ Office
❸ Staff WC
❹ Conference Room
❺ Equipment Room
❻ Activity Hall
❼ Balcony

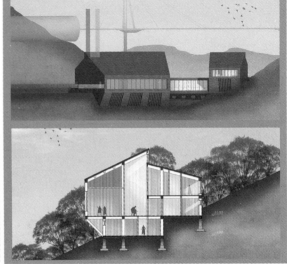

星光暗夜·山居秸院 4
Starlight Dark Night Hillhouse Straw Courtyard

· Room design strategy I

The monomer distribution of buildings has the characteristics of good privacy and strong terrain adaptability, but its shape coefficient is large, which is not conducive to heat preservation and insulation.

Cluster building distribution can reduce the shape coefficient of buildings and improve the thermal insulation performance of buildings, but its adaptability to terrain is poor.

After weighing the heat preservation and insulation performance of buildings and the adaptability of terrain, it is considered that the combined arrangement of guest rooms is more suitable for the site.

· Room design strategy II

· Determine the building shape.
· Increase the Sunshine.
· Optimize building Room, increase the shape basing on the index contact Aaea between seasonal solar height and angle.
· Increase south-facing outdoor landscape terrace combined with privacy and architectural form.

· Room design

· Aapartment A
Topographic Direction: Northeast Direction
Topographic gradient: 10%-20%
Floor Area: 41.58 ㎡ ×2

· Ground Floor Plan 1:100 ▲
· Second Floor Plan 1:100 ▶

· Aapartment B
Topographic Direction: Northeast Direction
Topographic gradient: 15%-30%
Floor Area: 41.58 ㎡ ×2

· Ground Floor Plan 1:100 ▲
· Second Floor Plan 1:100 ▶

A West Elevation 1:200 A South Elevation 1:200 B South Elevation 1:200 B East Elevation 1:200

· Aapartment C
Topographic Direction: Southwest Direction
Topographic gradient: 15%-30%
Floor Area: 55.44 ㎡ ×2

◀ · Ground Floor Plan 1:100
▼ · Second Floor Plan 1:100

· Aapartment D
Topographic Direction: Southwest Direction
Topographic gradient: 10%-20%
Floor Area: 56.7 ㎡ ×2

◀ · Ground Floor Plan 1:100
▼ · Second Floor Plan 1:100

C West Elevation 1:200 C South Elevation 1:200 D South Elevation 1:200 D East Elevation 1:200

· 1-1 Section 1:100

· 2-2 Section 1:100

According to the gradient distribution law of the northward slope of the terrain, two building forms are proposed to adapt to the two gradient ranges of the terrain. In order to maximize the lighting effect of the room and take into account the privacy, the traffic corridor is set on the north side of the building in the form of overhead. Because of the poor orientation, the room area is smaller.

· 3-3 Section 1:100

· 4-4 Section 1:100

Similarly, according to the gradient distribution law of the south slope of the terrain, two building forms are proposed to adapt to the two gradient ranges of the terrain. In order to maximize the lighting effect of the room and take into account the privacy, the traffic corridor is set in the north of the building in the form of semi-buried, cause of the better orientation, the room area is larger.

星光暗夜·山居秸院 5
Starlight Dark Night Hillhouse Strāw Courtyard

·Daylight factor analysis ·Ventilation analysis in summer

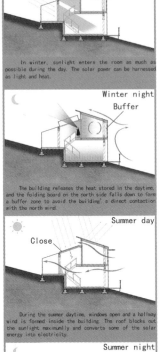

● Passive-form building design

Winter day

In winter, sunlight enters the room as much as possible during the day. The solar power can be harnessed as light and heat.

Winter night

Buffer

The building releases the heat stored in the daytime, and the folding board on the north side falls down to form a buffer zone to avoid the building's direct contaction with the north wind.

Summer day

Close

During the summer daytime, windows open and a hallway wind is formed inside the building. The roof blocks out the sunlight maximumlly and converts some of the solar energy into electricity.

Summer night

On summer nights, the hall wind passes through the building, the sun shade opens, and the building has the best viewing space.

Apartment A daylight factor | Daytime-Apartment A | Night-Apartment A
 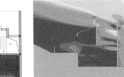 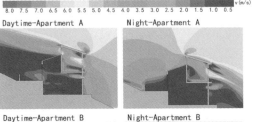

Apartment B daylight factor | Daytime-Apartment B | Night-Apartment B

Apartment C daylight factor | Daytime-Apartment C | Night-Apartment C
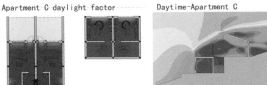 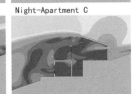 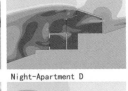

Apartment D daylight factor | Daytime-Apartment D | Night-Apartment D
   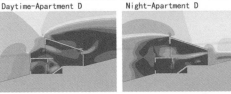

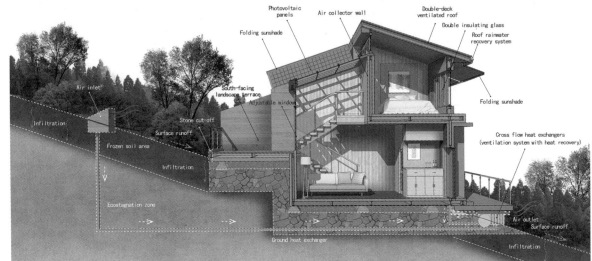

Labels: Photovoltaic panels; Air collector wall; Double-deck ventilated roof; Folding sunshade; Double insulating glass; Roof rainwater recovery system; Air inlet; South-facing landscape terrace; Adjustable window; Stone cut-off; Surface runoff; Frozen soil area; Folding sunshade; Infiltration; Cross flow heat exchangers (ventilation system with heat recovery); Ecostagnation zone; Air outlet; Surface runoff; Ground heat exchanger; Infiltration

星光暗夜·山居秸院 6
Starlight Dark Night Hillhouse Straw Courtyard

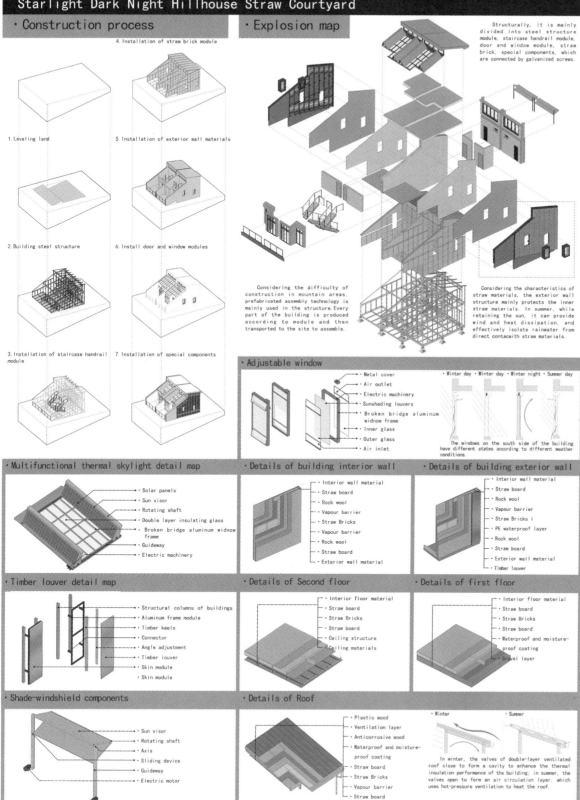

01 Relay Disappear & Stars Appear 驿隐星升

综合奖·优秀奖
General Prize Awarded·
Honorable Mention Prize

注 册 号：6003
项目名称：驿隐星升（兴隆）
　　　　　Relay Disappear & Stars Appear（Xinglong）
作　　者：付彤雨、吕 梦、崔向国
参赛单位：南京工业大学
指导老师：罗　靖

Geographic Location

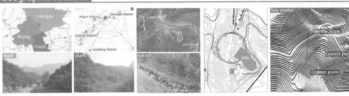

Concept Generation

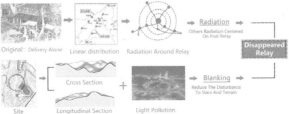

Map Relationship

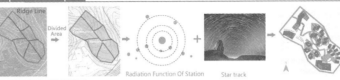

Design Explanation

本设计的主要理念为消隐的驿站，结合古代驿站体系的发展历程，提取辐射组团的元素，形成散点式布局。又从材料的消隐、建筑形体的消隐以及光的消隐出发，采用当地含量丰富的夯土资源作为和与山地较好融合的木材作为主要的建筑材料，利用太阳能光伏板技术与坡屋顶的结合，又顺应地势形成居住单元的单坡顶和公建的折屋顶。利用地形高差组织流线和单体建筑的垂直布局，利用可动格栅和特殊的室外灯光布置营造了良好的观星体验。

The main idea of this design is to eliminate the hidden station, combine the development process of the ancient station system, extract the elements of the radiation group, and form a scatter layout. Starting from the blanking of materials, the blanking of architectural forms and the disappearance of light, the local rich bauxite resources are used as the main building materials for the integration of wood with the mountain, using solar photovoltaic technology and slope roof. The combination of the single slope top and the publicly-built folding roof of the residential unit conforms to the terrain and uses the topographic height difference to organize the streamline and the vertical layout of the single building. A good star-gazing experience is created with a movable grille and special outdoor lighting lay.

Chengde District Analysis

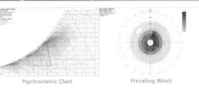

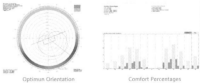

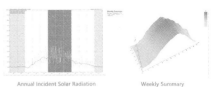

Site Climate Analysis

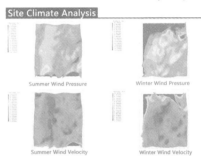

02 Relay Disappear & Stars Appear 驿隐星升

2019 台达杯国际太阳能建筑设计竞赛获奖作品集

Site Plan 1:500

Economic Technology Index
Covered Area : 7927 m²
Construction Area : 2446 m²
Building Density : 21.7%
Volume Rate : 0.31

天文台

Steel Grating Walkway Analysis

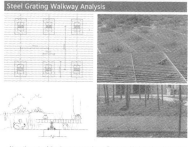

Along the route of the site, we arranged a small concrete block as a foundation with a stainless steel metal grille as a walkway. Such permeable walkways still allow plants to grow between them, and the activities of small animals are not interrupted by roads.

Site Climate Analysis (with architecture)

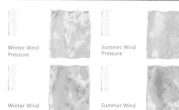

Winter Wind Pressure | Summer Wind Pressure
Winter Wind Velocity | Summer Wind Velocity

Regional Ecotect Analysis

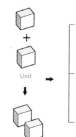

Family Residence Lighting | Sunshine Residence Lighting | Shadow Residence Lighting | Family Residence Ventilation | Sunshine Residence Ventilation | Shadow Residence Ventilation

Block Generation Analysis

Residential Unit Analysis

Unit → Group

- Embedded → Height Adjustment → **Shadow Residence**
- Separation → Height Adjustment → **Sunshine Residence**
- Wndraw → Height Adjustment → **Family Residence**

Reception Hall Analysis

Volume → Deformation → Height Adjustment → Reception Hall

Roof Slope Analysis

Obscured | Unobstructed

Solar radiation angle in winter : 70°
Solar radiation angle in summer : 10°

Heat Preservation In Winter
Energy Supply In Summer

- Choose 10° as the slope
- Increase the snowproof eaves

snow eaves

Restaurant Analysis

Plan → Block + Terrain → Slope Top → Final Version → Restaurant

173

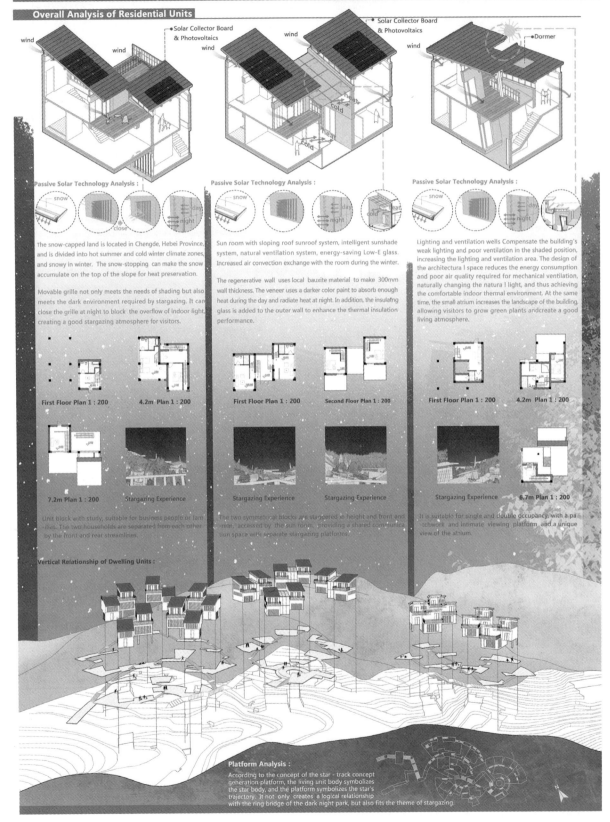

04 Relay Disappear & Stars Appear 驿隐星升

Overall Analysis of the Restaurant Section

① The Hall
② Dining Area
③ The Kitchen
④ Laundey Room
⑤ Staff Louge
⑥ Staff Office
⑦ Equipment Room
⑧ Male Toilet
⑨ Female Toilet

First Floor Plan 1 : 200

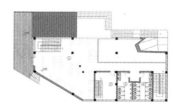

① Dining Area
② Meal Delivery Room
③ Male Toilet
④ Female Toilet
⑤ Outdoor Platform

Second Floor Plan 1 : 200

Northwest Facade 1 : 200

Northeast Facade 1 : 200

The Application of Solar Technology in Restaurants:

Active Solar Technology: Technology uses solar panels to generate electricity, and the energy needs of the restaurant itself can also store excess energy. Solar collectors and ground source heat pumps are heating in cold. The rainwater harvesting system recovers resources through organized drainage on the roof.

Passive Solar Technology: It also reflects the passive solar technology used in residential units. Regeneration walls and air buckets are used for keep warm, movable grilles are used for shading and lighting, and top windows are used for ventilation. The second layer of the platform realizes the communication connection between the station and the observatory.

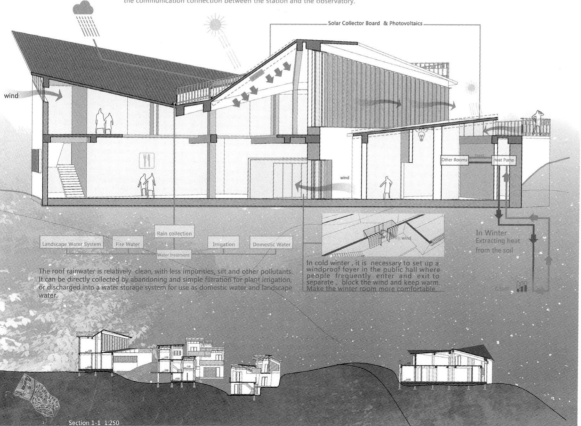

The roof rainwater is relatively clean, with less impurities, silt and other pollutants. It can be directly collected by abandoning and simple filtration for plant irrigation, or discharged into a water storage system for use as domestic water and landscape water.

In cold winter, it is necessary to set up a windproof foyer in the public hall where people frequently enter and exit to separate, block the wind and keep warm. Make the winter room more comfortable.

Section 1-1 1:250

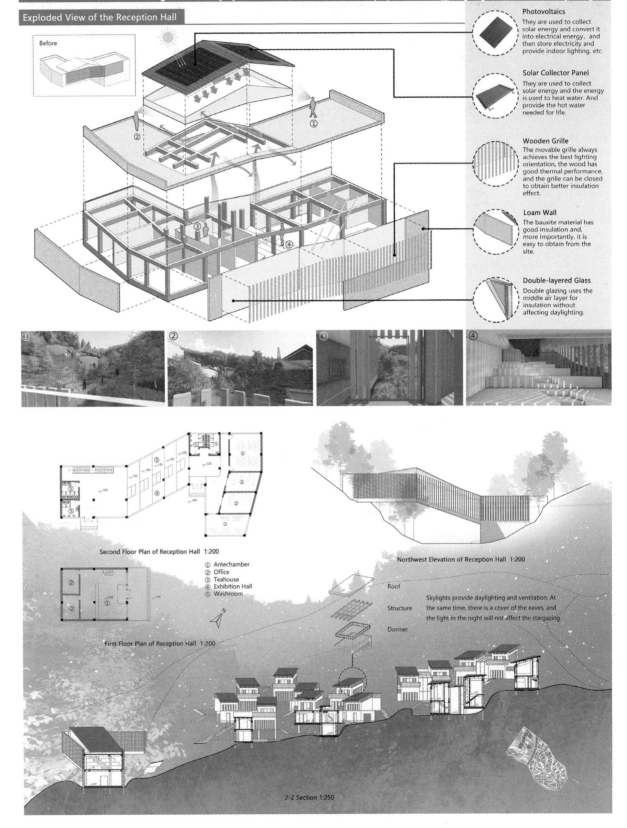

06 Relay Disappear & Stars Appear 驿隐星升

The protruding eaves can block the light that hits the sky, preventing the diffused light from interfering with the stargazing.

The movable grille can adjust the direction of light illumination to avoid affecting the stargazing activity outside the room.

Set street lights under the armrests to ensure that there is no light source interference in the field of view when people look up at the stars.

Set lighting under the building, use the building to block the diffuse light towards the sky, optimize the light environment.

Exploded View of Café

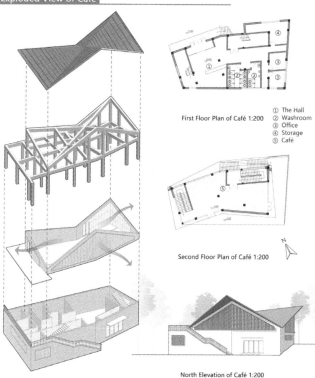

First Floor Plan of Café 1:200
① The Hall
② Washroom
③ Office
④ Storage
⑤ Café

Second Floor Plan of Café 1:200

North Elevation of Café 1:200

Visual Field Analysis

The height difference at the entrance of the site is obvious. The hallway is used to limit the view, and the scene inside the site is almost invisible.

After entering the site, the field of view gradually opens, but the direction is obvious due to the linearly distributed buildings on both sides

Eventually came to the inner circular plaza, the buildings on both sides retreat, the horizon is instantly open

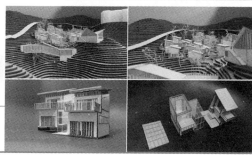

01 光谷星桥 SUN VALLEY · STAR BRIDGE

综合奖・优秀奖
General Prize Awarded ·
Honorable Mention Prize

注 册 号：6016
项目名称：光谷星桥（兴隆）
　　　　　Sun Valley · Star Bridge
　　　　　(Xinglong)
作　　者：周春妮、陈晗、程吉帆、
　　　　　陈胜蓝、曾港俊
参赛单位：宙思建筑设计（上海）事务所

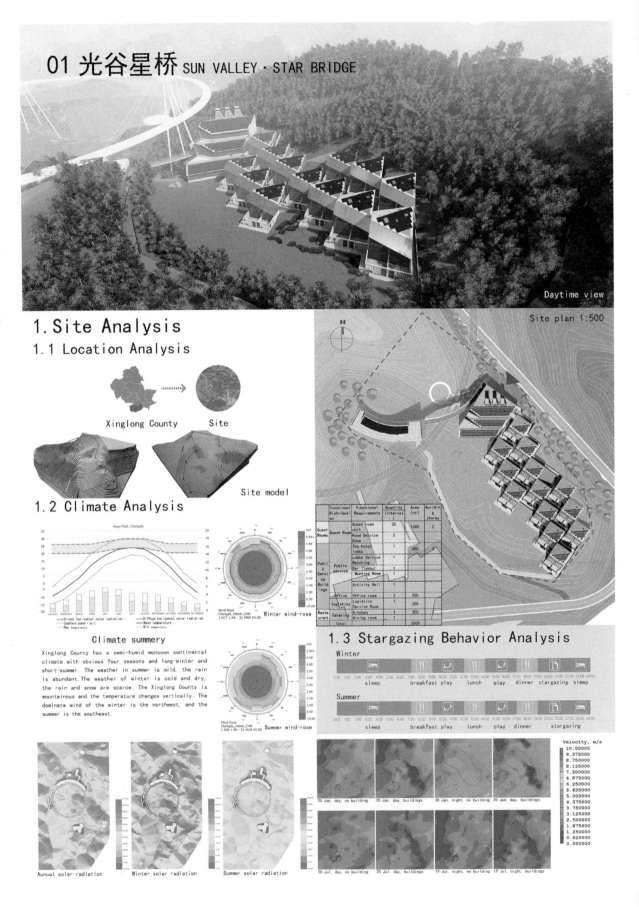

1. Site Analysis
1.1 Location Analysis
1.2 Climate Analysis

Climate summery

Xinglong County has a semi-humid monsoon continental climate with obvious four seasons and long-winter and short-summer. The weather in summer is mild, the rain is abundant. The weather of winter is cold and dry, the rain and snow are scarce. The Xinglong County is mountainous and the temperature changes vertically. The dominate wind of the winter is the northwest, and the summer is the southeast.

1.3 Stargazing Behavior Analysis

02 光谷星桥 SUN VALLEY · STAR BRIDGE

设计说明

本项目以"光谷星桥"为主题，针对山谷地势以及寒冷地区冬长夏短的特点，以获得南向太阳辐射且冬季防风、夏季通风作为规划依据，公共服务建筑和餐厅布置在上风向，客房部分在下风向，各部分以栈桥连接。在被动式技术上，客房部分依山就势，南向退台，设置阳光房、solar envelope 屋顶、特伦布墙和活动式院墙为观星活动提供舒适环境；公共服务部分采用面朝南向阶梯式布局，办公等常用公共空间设置在南侧，以活动式水平遮阳保持舒适照度，中庭通风采光井设置拔风塔，餐厅以厨房作为热源，结合 GR 相变材料，将厨房余热传递至餐厅空间。在主动式技术上，采用太阳能光伏发电、雨水收集、水源热泵和热回收系统以实现低能耗建筑的目标。

Design Description

Sun Valley and Star Bridge, based on the valley topography and the weather of long-winter and short-summer in cold regions, is aimed to obtain southward solar radiation, winter windproof and summer ventilation. The public service building and restaurants are arranged in the upwind direction, and the guest rooms are settled in the downwind area. Those three parts are connected by a bridge in the valley. In the passive technology, the guest rooms are going up the mountains and retreating toward south. The transitional space, solar envelope roof, the Trombe wall and the movable courtyard grills provide a comfortable environment for stargazing activities. The public service building adopts a south-facing ladder layout, office and other common public spaces are set on the south side, with the movable horizontal shading devices to maintain comfortable illumination. The atrium and the wind tower are good for ventilation and lighting. The kitchen is used as a heat source in the restaurant, combined with the GR phase change material to keep the heat, it transfer the excess heat to the restaurant space. In active technology, solar photovoltaic power generation, rainwater collection, water source heat pumps and heat recovery system are used to achieve the goal of low energy buildings.

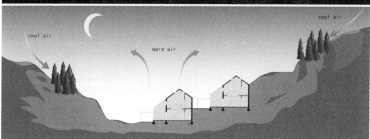

Nightscape

2. Planning

2.1 Earthwork and Orientation Selection

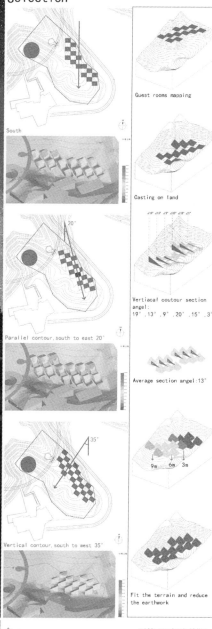

South

Parallel contour, south to east 20°

Vertical contour section angle: 19°, 13°, 9°, 20°, 15°, 3°

Average section angle: 13°

Vertical contour, south to west 35°

Guest rooms mapping

Casting on land

Fit the terrain and reduce the earthwork

2.2 Valley Wind Diagram

Set the guest rooms on the northeast slope to get more sunshine. plant trees in the upwind direction of the valley upper part at night to reduce the valley wind speed.

2.3 Site Planning Diagram

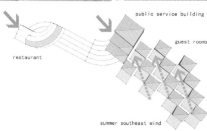

The idea of the overall planning layout is to use the tall public buildings to block the northwest wind in winter, and the low-rise guest rooms are placed on the northeast slopes.

03 光谷星桥 SUN VALLEY · STAR BRIDGE

Living room | Bedroom

3.1 Diagram

original volume | solar envelope 33° | best solar PV angle 36° | transitional space (solar house)

3. Guest Rooms Design

The guest room part is made up of 13 villas, which are mainly divided into two parts: interior duplex two-story house and outdoor courtyard. The sloping roof extends southward to increase the utilization ratio of solar PV panels, while the solar house glass facade still brings more light and heat to the interior. The southwest wall is equipped with a trombe wall to heat air into the room. The south facing roof panel adopts an inclination angle of 36° as the best solar panel power generation. The northward roof uses the principle of solar envelope, which does not obstruct the sunshine of rear row of buildings.

3.2 Solar House Comparison

Heating set temp:18°C
Time: from Oct to Mar

Case1: open to sky heating load=84kWh/m²
Case2: cover (H=2m) heating load=53kWh/m²
Case3: cover (H=3m) heating load = 54 kWh/m²

Case4: cover (H=4m) heating load=55kWh/m²
Case5: slope cover (H=2~3m) heating load=49kWh/m²
Case6: slope cover (H=2~4m) heating load=46kWh/m²
Case7: cover (H=3m) heating load=38kWh/m²

Guest room envelop

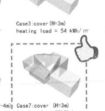

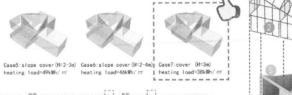

W/W Ratio Southwest : 0.3 | W/W Ratio Southeast : 0.8 | W/W Ratio Skylight : 0.1

Ventilation mode: The outside air temperature is naturally ventilated during 18~26 °C, and the window is closed when the outside temperature is less than 18 °C or higher than 26 °C.
Usage mode: 2 users, use the room from 22:00pm to 7:00am. The main electrical appliances are refrigerators, electric kettles, water heaters, and electric lights.
Air conditioning mode: WSHP + DOAS water source heat pump, heat recovery fresh air system.

① Sloping roof

Photovoltaic panels
Asphalt waterproof layer
Thermal insulation layer 100mm
Concrete slab layer 50mm
Cement mortar layer 50mm
Wood plate surface 10mm

② Double glazing

Multiple glazings
Low-E coating
Gas fill
Warm edge spacer

③ Triple glazing

Multiple glazings
Low-E coating
Gas fill
Warm edge spacer

④ Trombe wall
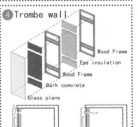
Wood Frame
Eps insulation
Wood Frame
Dark concrete
Glass plane
Summer | Winter

⑤ Northwest wall

Wood plate surface 10mm
Cement mortar layer 50mm
Thermal insulation layer 100mm
Concrete slab layer 200mm
Cement mortar layer 50mm

⑥ Southeast wall
Wood plate surface 10mm
Cement mortar layer 50mm
Thermal insulation layer 100mm
Concrete slab layer 100mm
Cement mortar layer 50mm

⑤ The courtyard
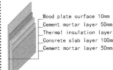
Summer | Winter

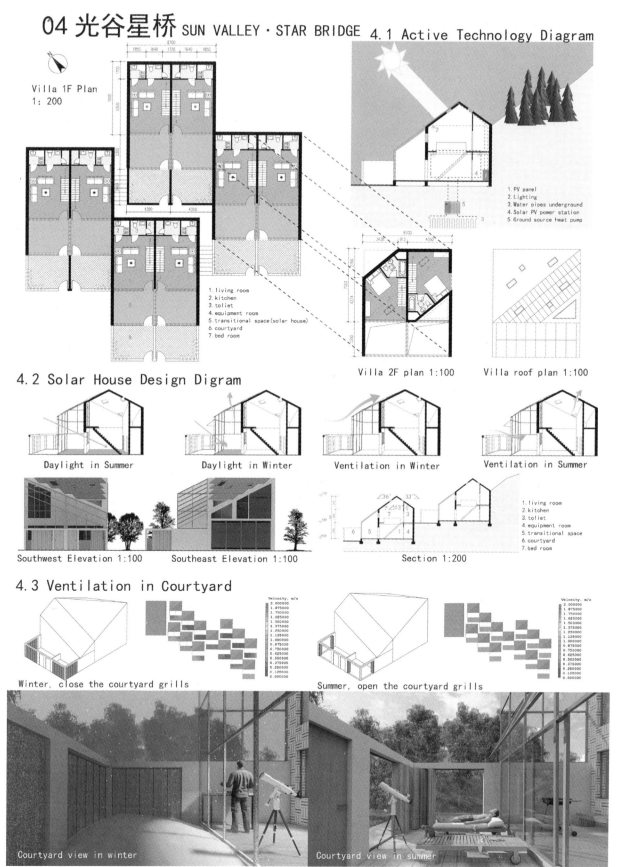

05 光谷星桥 SUN VALLEY · STAR BRIDGE

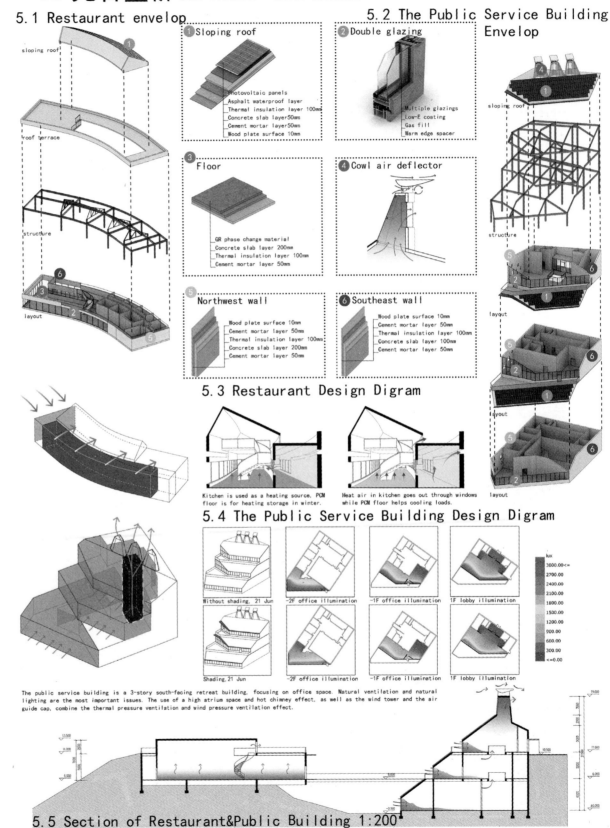

The public service building is a 3-story south-facing retreat building, focusing on office space. Natural ventilation and natural lighting are the most important issues. The use of a high atrium space and hot chimney effect, as well as the wind tower and the air guide cap, combine the thermal pressure ventilation and wind pressure ventilation effect.

5.5 Section of Restaurant & Public Building 1:200

06 光谷星桥 SUN VALLEY · STAR BRIDGE

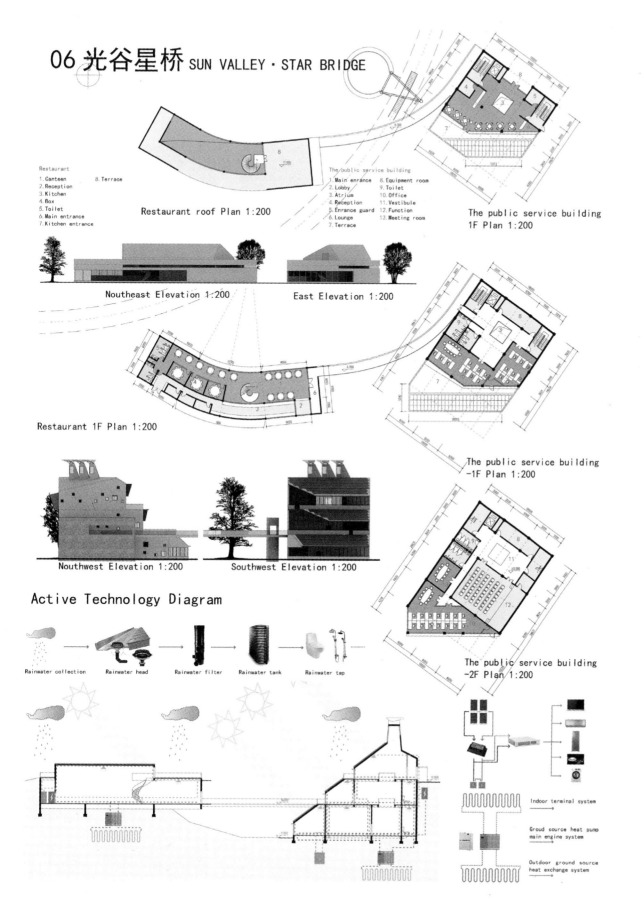

综合奖·优秀奖
General Prize Awarded · Honorable Mention Prize

注 册 号：6032
项目名称：山居（兴隆）
　　　　　Mountain Perch（Xinglong）
作　　者：王博达、田 佳、沈梦丹
参赛单位：华中科技大学、武汉理工大学
指导老师：徐 燊

SITE ANALYSIS

Chengde

Xinglong

SITE

PLANNING AREA

CLIMATE ANALYSIS

OPTIMUM ORIENTATION　　PSYCHROMETRIC CHART　　AVERAGE TEMPERATURE

WIND FREQUENCY　　AVERAGE WIND TEMPERATURES　　CLIMATIC CONDITIONS

设计说明

本方案以山间小屋为设计想法，场地布局和建筑设计均结合基地山谷的地形地貌，顺应基地等高线采用线状布局，形成上下错落的景观面；结合场地高差进行建筑内部空间设计，满足了酒店的功能需求。

规划和设计中充分考虑了当地的气候环境。规划布局考虑当地方向风，形成了基地内部的通风走廊；建筑设计中通过模拟分析当地的太阳辐射在建筑表面的分布，在适宜的表面安装太阳能光伏板来提供电力需求。此外，建筑设计中还考虑采用地源热泵系统，以解决冬季采暖和夏季降温，同时设置阳光间、保温墙体、双层玻璃等被动式节能技术，以达到舒适和节能的效果。

DESIGN DESCRIPTION

The scheme is based on the idea of a mountain hut, and the site layout and architectural design combine the topography of the base valley. Linear layout is adopted to conform to the contour line of the site to form a landscape surface that is upside down; the interior space design of the building is combined with the height difference of the site, which meets the functional requirements of the hotel.

The local climate is fully considered in planning and design. The planning layout considers the local wind direction and forms the ventilation corridor inside the site. In the architectural design, the distribution of local solar radiation on the building surface is simulated and the solar photovoltaic panels are installed on the appropriate surface to provide electricity demand. In addition, the ground source heat pump system is also considered in the architectural design to solve the problem of heating in winter and cooling in summer. At the same time, passive energy-saving technologies such as sun room, insulation wall and double-glazing are set up to achieve the effect of comfort and energy saving.

ANALYSIS OF SITE STATUS

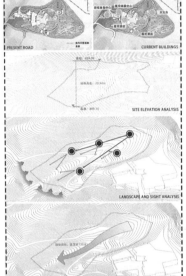

PRESENT ROAD　　CURRENT BUILDINGS

SITE ELEVATION ANALYSIS

LANDSCAPE AND SIGHT ANALYSIS

SITE MORPHOLOGICAL ANALYSIS

VISITOR ANALYSIS

ANALYSIS OF TOURIST BEHAVIOR

ANALYSIS OF STARGAZING ACTIVITIES

AERIAL VIEW

山居——Mountain Perch 01

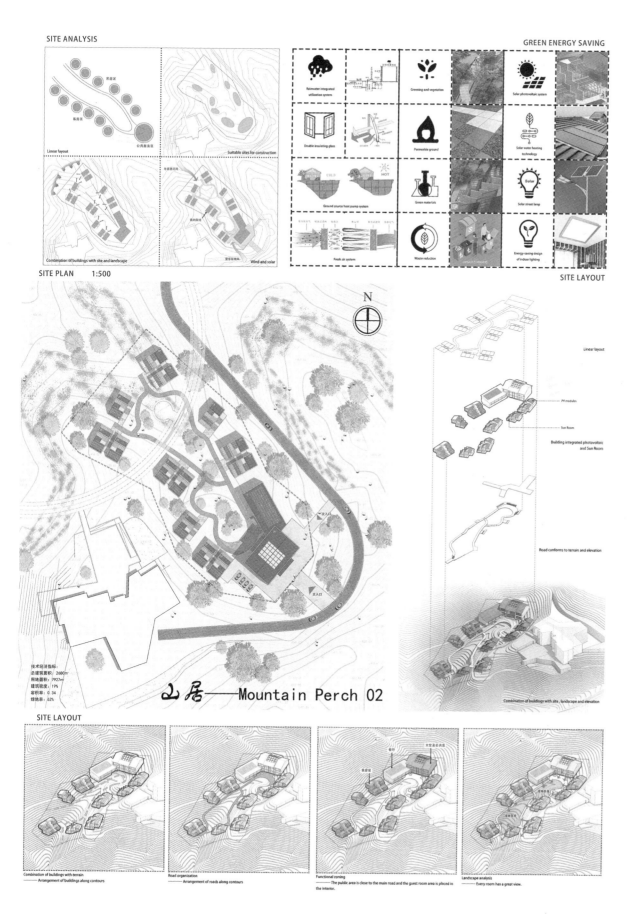

山居——Mountain Perch 03

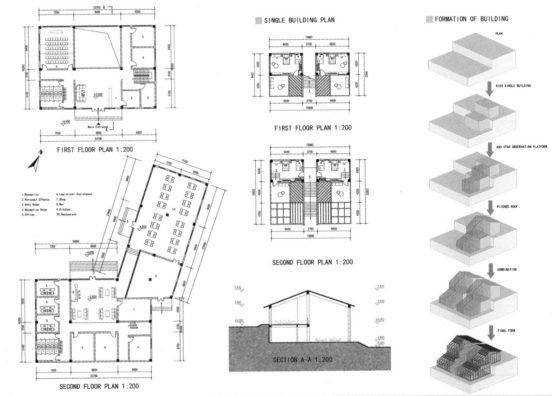

Formation of architectural forms with different slopes | Solar radiation analysis

2019 台达杯国际太阳能建筑设计竞赛获奖作品集

ACTIVE AND PASSIVE TECHNOLOGY INTENTION

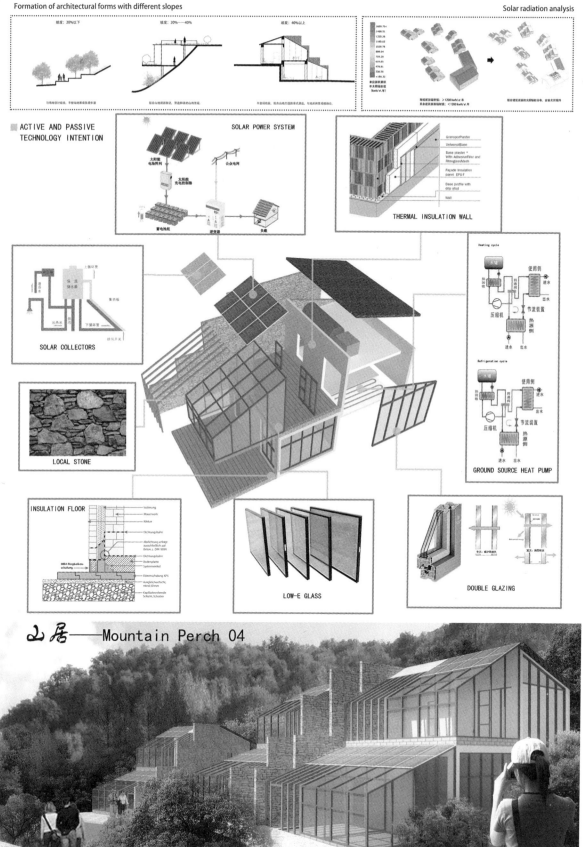

山居——Mountain Perch 04

综合奖·优秀奖
General Prize Awarded · Honorable Mention Prize

注册号：6044
项目名称：白驹临庭 繁星入梦（兴隆）
Sunshine Pouring in Courts
Stars Twinkin in Dream
（Xinglong）
作　者：程梦琪、洪叶、刘晶、范静哲、初楚、李佳宸、高铖
参赛单位：南昌大学
指导老师：叶雨辰、郭兴国

白驹临庭 繁星入梦 02
SUNSHINE POURING IN COURTS STARS TWINKLING IN DREAMS

2019 台达杯国际太阳能建筑设计竞赛获奖作品集

AIRSCAPE

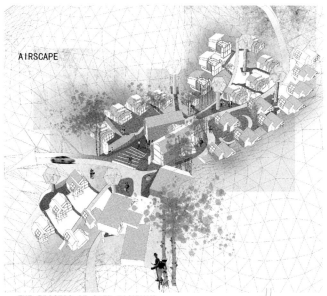

ARCHITECTURAL GENERATION PROCESS

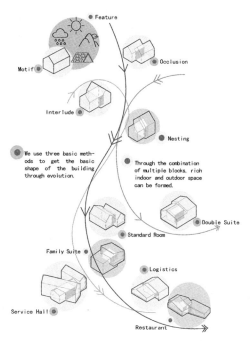

We use three basic methods to get the basic shape of the building through evolution.

Through the combination of multiple blocks, rich indoor and outdoor space can be formed.

THE PROCESS OF SITE PLANNING

1. Extraction of elements in VanGogh's famous works — "Star Moon Night".
2. Put elements into the site.
3. Divide area of sound.

4. Identify primary and secondary entrances.
5. Place public parts in downtown areas.
6. Analyzing the site conditions, the site will be zoned.

7. Different room units are placed in different areas.
8. Add trails to each room unit.
9. Add greening and public star-watching places.

PRIVACY STRATEG

Problem: The units are too close to each other.

Strategy1: Place buildings in the same direction and add grilles

Strategy2: Grow plants between buildings and beside roads

THE ANALYIS OF SITE HANDING

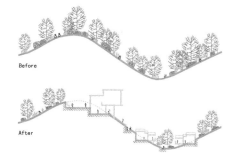

Before

After

SITE PLAN 1:500

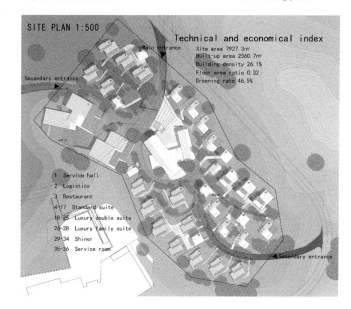

Technical and economical index
Site area 7927.3㎡
Built-up area 2560.7㎡
Building density 26.1%
Floor area ratio 0.32
Greening rate 46.5%

1 Service hall
2 Logistics
3 Restaurant
4-17 Standard suite
18-25 Luxury double suite
26-28 Luxury family suite
29-34 Shiner
35-36 Service room

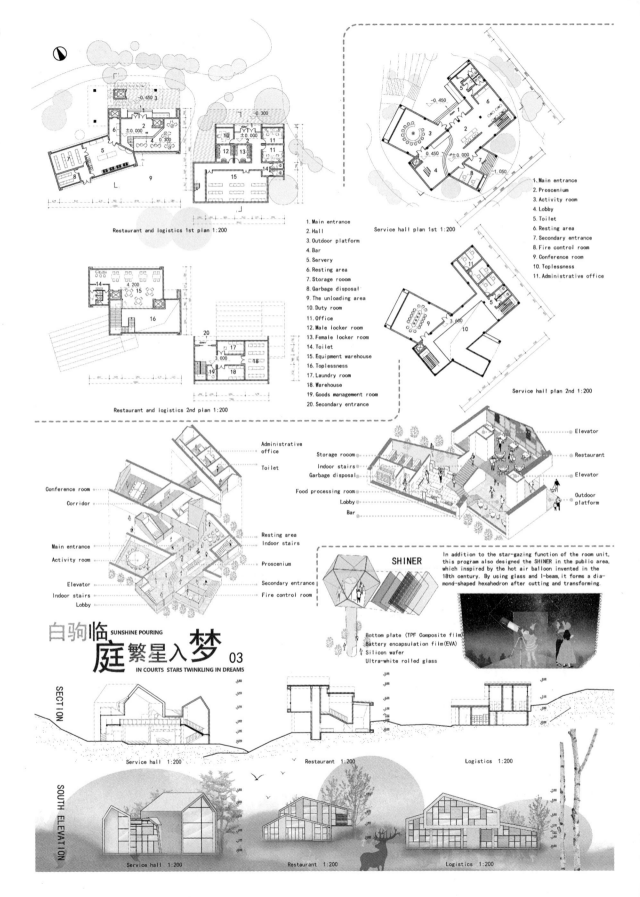

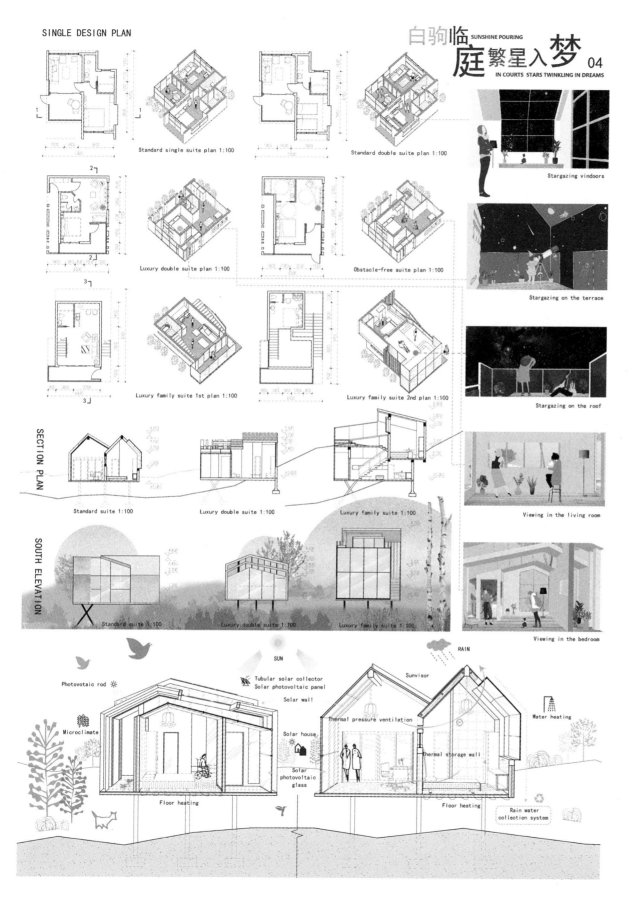

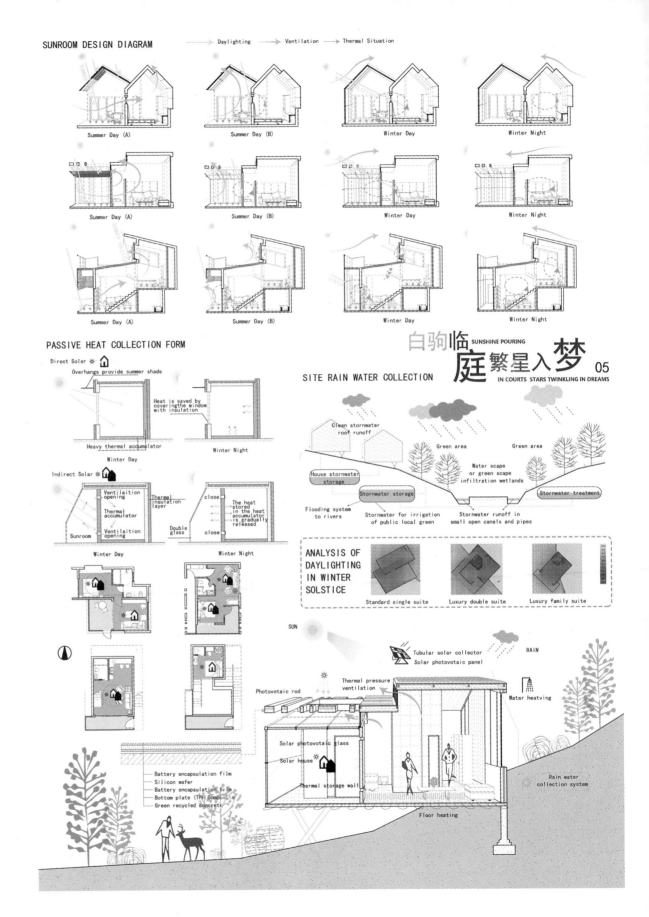

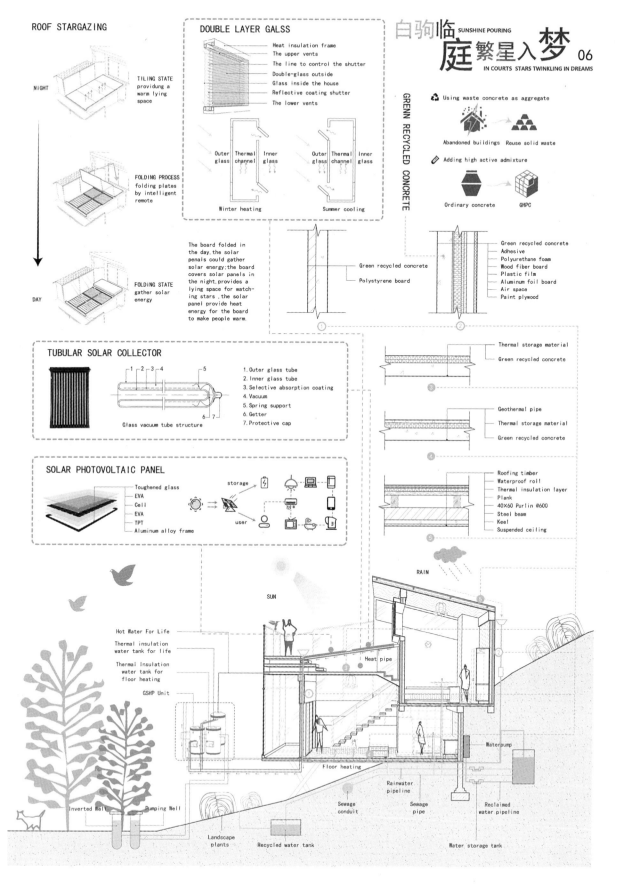

综合奖·优秀奖
General Prize Awarded · Honorable Mention Prize

注 册 号：6173
项目名称：归源山居（兴隆）
　　　　　Guiyuan Mountain Residence
　　　　　(Xinglong)
作　　者：赵谷橙、岳小超、齐梦晓、
　　　　　李志权、金宁园、王　晨
参赛单位：河北工业大学
指导老师：李有芳

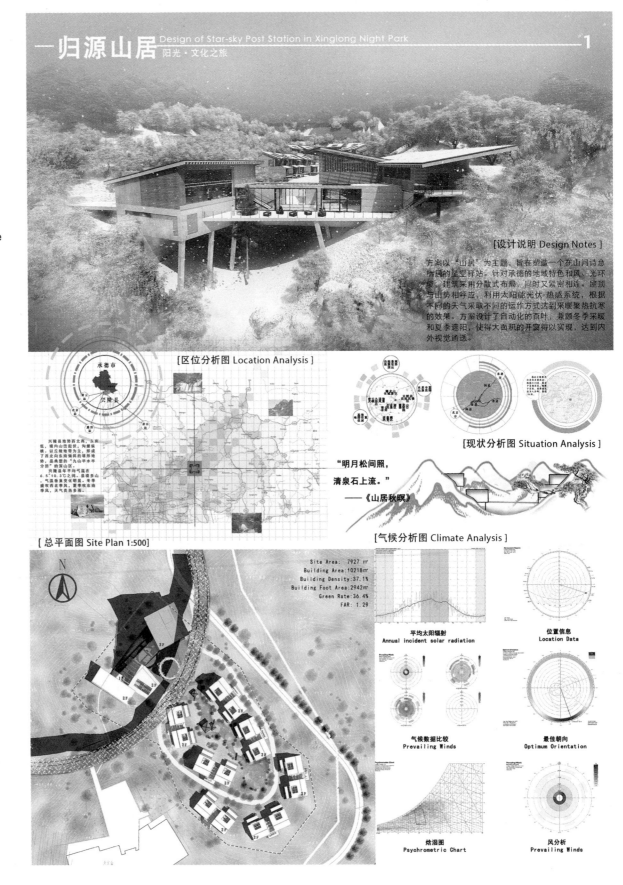

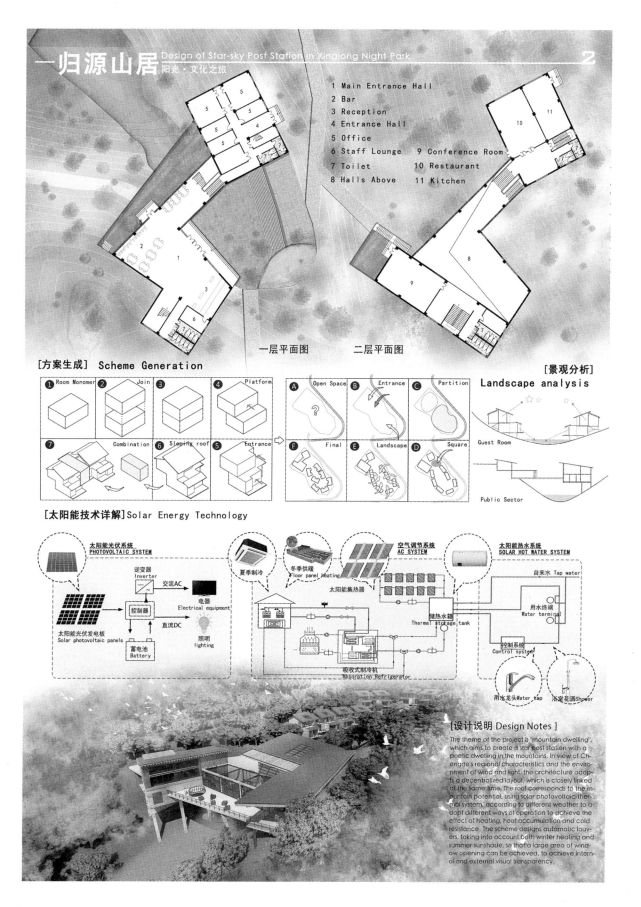

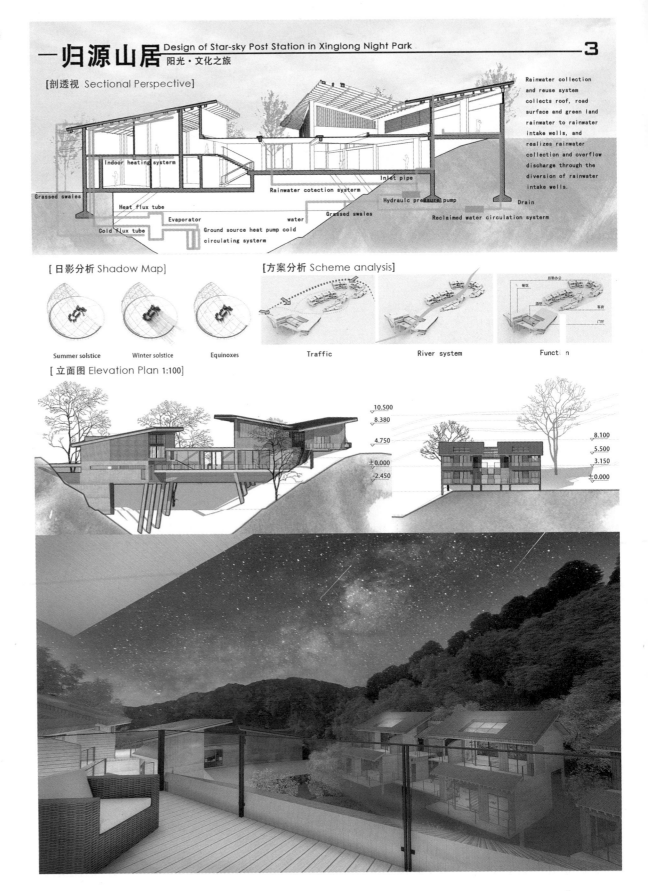

归源山居 Design of Star-sky Post Station in Xinglong Night Park
阳光·文化之旅

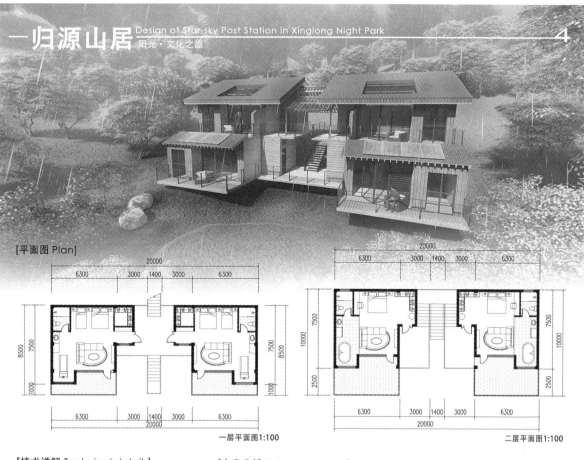

[平面图 Plan]

一层平面1:100　　　二层平面图1:100

[技术详解 Technical details]

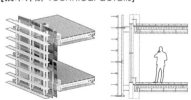

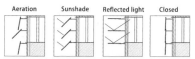

The sunshade is a semi-active epidermis, which can prevent residents from direct sunlight or isolation of sunlight in the indoor. The system also provides feasibility optimization of building services, such as heat pump control, because it can run automatically, heating the air behind the module in winter or introducing fresh and cool air from outside in summer.

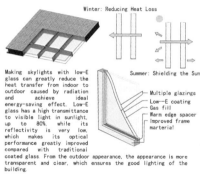

Making skylights with low-E glass can greatly reduce the heat transfer from indoor to outdoor caused by radiation and achieve ideal energy-saving effect. Low-E glass has a high transmittance to visible light in sunlight, up to 80%, while its reflectivity is very low, which makes its optical performance greatly improved compared with traditional coated glass. From the outdoor appearance, the appearance is more transparent and clear, which ensures the good lighting of the building.

[方案分析 Scheme analysis]

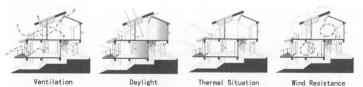

Ventilation　　Daylight　　Thermal Situation　　Wind Resistance

[剖透视 Anatomical Perspective]

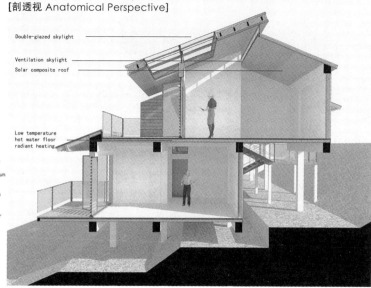

综合奖·优秀奖
General Prize Awarded·Honorable Mention Prize

注 册 号：6183
项目名称：星之所向（兴隆）
　　　　　Direction of Stars（Xinglong）
作　　者：石 丹、耿艺曼
参赛单位：重庆大学
指导老师：周铁军、张海滨

星之所向　The direction of stars

区位分析

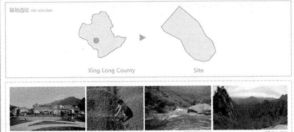

基地概况 Base survey

兴隆县，地处河北省东北部，承德市最南端，长城北侧。兴隆县地势西北高，东南低，境内山峦起伏，沟壑纵横。以丘陵地带为主，形成了西北向东南倾斜的塔形地势，是典型的"九山半水半分田"的深山区。兴隆县年平均气温在6.5~10.3℃之间。县境多山，气温垂直变化明显。兴隆县是全国的"山楂之乡"、"板栗之乡"。拥有雾灵山、六里坪森林公园、兴隆溶洞北京天文台兴隆观测站等著名景点。

Xinglong County is located in the northeast of Hebei Province, the southernmost end of Chengde City and the north side of the Great Wall. Xinglong County is high in the northwest and low in the southeast. The hills are rolling and the gullies are vertical and horizontal. With hilly areas as the main area, a tower-shaped terrain inclined northwest to southeast has been formed, which is a typical "Nine" Mountains, waters and fields are divided in the deep mountain areas. The average annual temperature in Xinglong County ranges from 6.5 to 10.3 degrees Celsius. The county is mountainous and the temperature changes vertically Obvious. Xinglong County is the "Hawthorn Township" and "Chestnut Township" of the whole country. Having Xinglong Karst Cave in Liuliping Forest Park, Wuling Mountain Xinglong Observatory of Beijing Observatory and other famous scenic spots.

场地气候分析

星之所向   *The direction of stars* 2/4

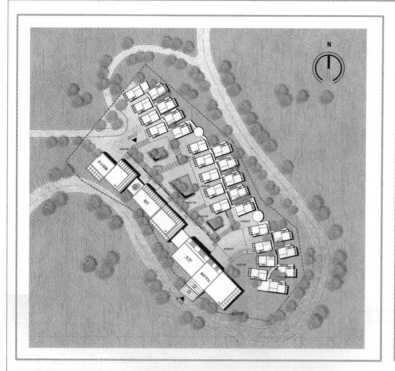

总平面图 1:1000

应对策略 Coping strategies

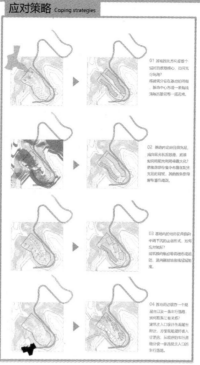

设计说明：

项目用地位于河北省承德市兴隆县六道河镇安营泰村暗夜公园（拟划中）内。暗夜公园处于山谷两侧的坡地上，主要建筑用房为25个单体客房以及餐厅大堂等服务型用房，我们在设计时根据地形、气候以及日照的限制条件将客房布置在北侧，服务型用房置于南侧，争取阳光的最大化利用。客房采用装配式技术，整体吊装，并在建筑中设置太阳能板、风幡等节能技术手段实现建筑的绿色节能目的。

The project site is located in the night peak of Anying Village, Liudaohe Town, Xinglong County, Chengde City, Hebei Province (in planning) Inside. Night Park is located on the slopes on both sides of the valley. The main building rooms are 25 single rooms.Service rooms such as restaurant lobby are designed according to terrain, climate and sunshine restrictions.Conditions will be arranged in the north side of the room, service-oriented rooms in the South side, strive for the maximum use of sunshine.The guest rooms adopt assembly technology and are lifted as a whole, and solar panels and wind caps are installed in the buildings to save energy. Technological means to achieve the goal of green energy saving in buildings.

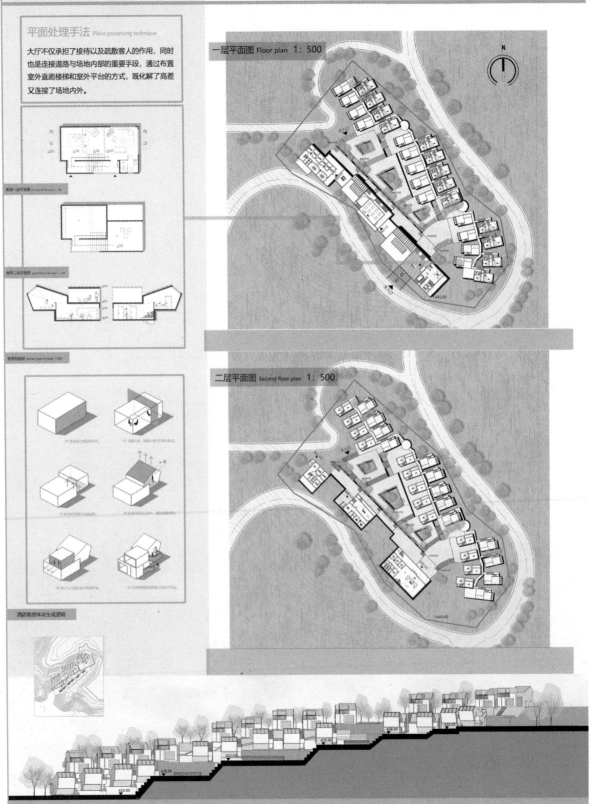

星之所同 — The direction of stars 4/4

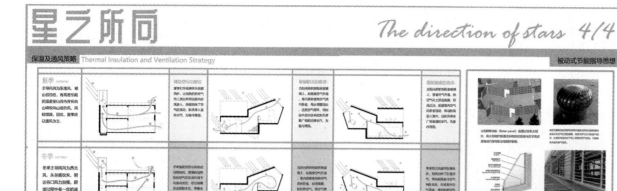

保温及通风策略 / Thermal Insulation and Ventilation Strategy

被动式节能指导思想

雨水收集系统 / Rainwater collection system

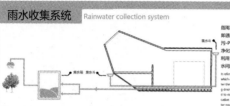

Biogas collection system

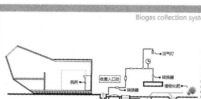

客厅装配式构造图 / Assembly drawing of living room

保温隔热技术 / Hotel Room Block Generation Strategy

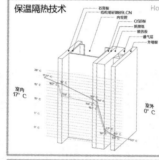

轻钢结构的墙体构造有内、外双层空腔，冬天让双层空腔封闭，空腔内的空气就不会流通，起到双层空气热阻的作用，能达到很好的保温效果（1个内空腔可产生3.9℃温差，节能约40%）。夏天，让内、外双层空腔上下开通，利用热空气往上走的热压通风原理，让热空气通过空腔排到屋顶外，达到节能的效果。

客房装配式构造图 / Guest room assembly drawing
卫生间装配式构造图 / Toilet assembly drawing

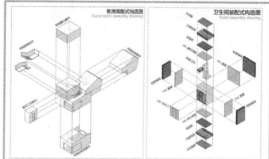

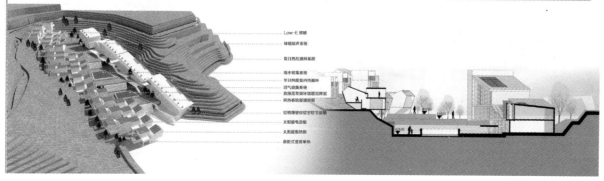

综合奖·优秀奖
General Prize Awarded · Honorable Mention Prize

注 册 号：6186
项目名称：落星谷（兴隆）
　　　　　The Valley of Falling Star (Xinglong)
作　　者：江海华、张 晨、谢 楠、宣姝颖
参赛单位：华中科技大学
指导老师：徐 燊

落星谷——河北兴隆暗夜公园星空驿站设计

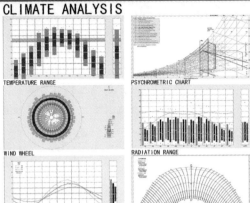

设计说明

本设计以星座为主题，在充分考虑山谷地理条件之后，将星象与总平面布局点结合。在地面勾勒出两个常见的星座——北斗七星和仙后座，并据此确定北极星的位置。天上星与地上星相呼应，在山谷中重现了一幅梦幻的星象图。

综合考虑当地地理气候和环境，采用主被动式结合的方式减少能耗。主要利用阳光房和种植屋面调节建筑微环境，并集主动式太阳能、沼气、地源热泵实现以替换供给能量并循环使用，同时从整体规划考虑太阳能、沼气等能源的共享。

DESIGN DESCRIPTION

This design takes "the constellations" as the theme. After fully considering the geographical conditions of the valley, We combined the star and the general layout point. Two familiar constellations are outlined on the ground — Big Dipper and Cassiopeia, and they determine the location of the north star. The stars on the ground echo the stars in the sky, recreating a dreamlike star map in the valley.

Considering local geography, climate and environment, Reduce energy consumption by combining active and passive methods. The sun room and planting roof are mainly used to adjust the architectural microenvironment. The combination of active solar, biogas, ground source heat pump is realized to replace the energy supply and recycle. Also, the sharing of energy such as solar/biogas is considered in the overall planning.

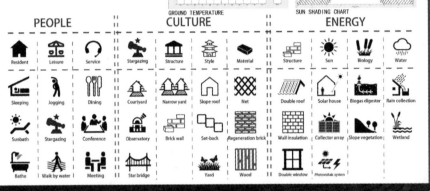

THE VALLEY OF FALLING STARS　01

落星谷——河北兴隆暗夜公园星空驿站设计

CONCEPT OF CONSTELLATION

The project adopts the method of putting the site first.

According to the analysis of the site, the buildings with appropriate volumes are arranged at the most appropriate points. The selected points are connected by roads and earth lamps to form the two constellations most commonly used by people who are just beginning to learn stargazing. The location of the Pole Star is thus determined, where the restaurant is located.

The scheme is not only suitable for the physical conditions of the site, but also echoes the star theme.

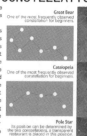

Great Bear — One of the most frequently observed constellation for beginners.

Cassiopeia — One of the most frequently observed constellation for beginners.

Pole Star — Its position can be determined by the two constellations, a transparent restaurant is placed in this position.

GENERATION OF LAYOUT

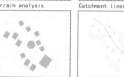

Terrain analysis | Catchment lines | Sunshine analysis | Wind analysis

The buildings are arranged in relatively flat positions. | After construction, the catchment is more concentrated. | Arrange sunshine room to the sunny side. | Introduce summer air into rooms to form a draft, and isolate the winter wind.

SITE PLAN 1:500

ECONOMIC INDEX
TOTAL LAND AREA: 7950 m²
TOTAL CONSTRUCTION AREA: 2510 m²
FAR: 31.6%
FIRST FLOOR RATIO: 26.4%
LANDSCAPE RATIO: 60.7%

LANDSCAPE ANALYSIS

The project emphasizes the development of low impact, the area is mainly on foot, the wooden trestle board, the wooden platform, the soft net and the hard net which can be contacted with water and the plants, and the rock dam are set up to form the different travel experience. The relationship between man and nature has always been "within sight but beyond destruction".

SOFT NET+WATER | SOFT NET+PLANTS | HARD NET+WATER | WOODEN PLATFORM+SLOPE | DAM+WATER

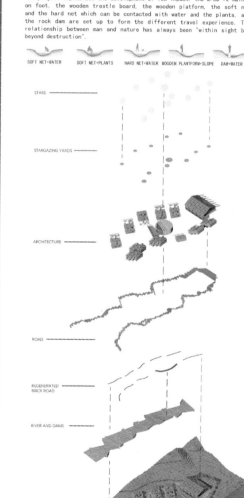

STARS
STARGAZING YARDS
ARCHITECTURE
ROAD
REGENERATED BRICK ROAD
RIVER AND DAMS

MATERIALS IN LANDSCAPE

REGENERATION OF BRICK — It is mainly made of building debris from nearby areas. As paving, it also plays a role in increasing water seepage.

WOOD — The wooden walkway takes advantage of the abundance of local wood.

GREY TILE — The dam is made of recycled tiles, which enriches the facade.

GRAVEL — The gravel at the bottom of the river also comes from recycled construction waste. People can step on the riverbed during the dry season.

HEMP ROPE NET — The net is placed over the river and grassland, which is convenient for people to get close to nature.

THE VALLEY OF FALLING STARS

02

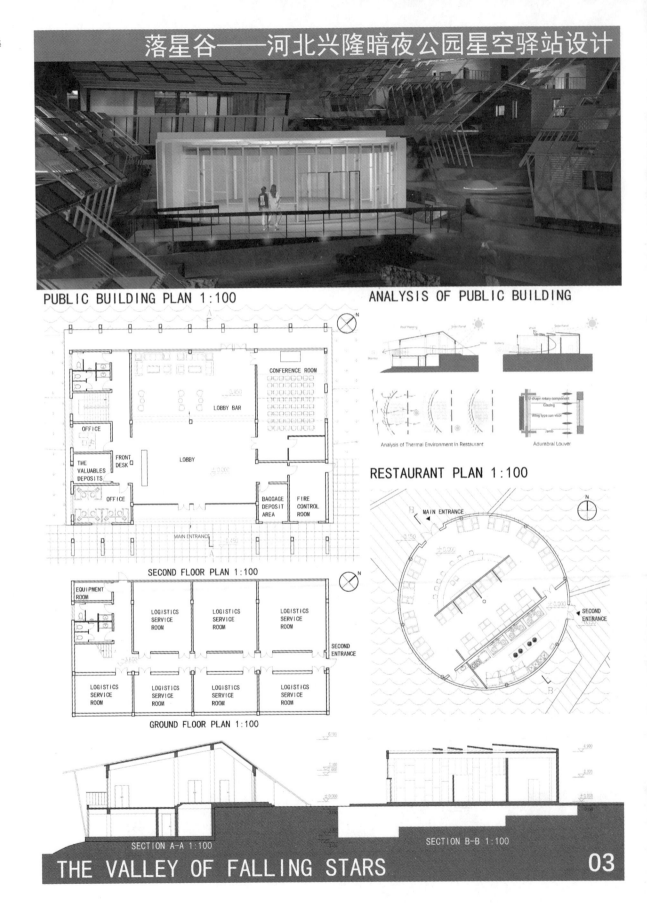

落星谷——河北兴隆暗夜公园星空驿站设计

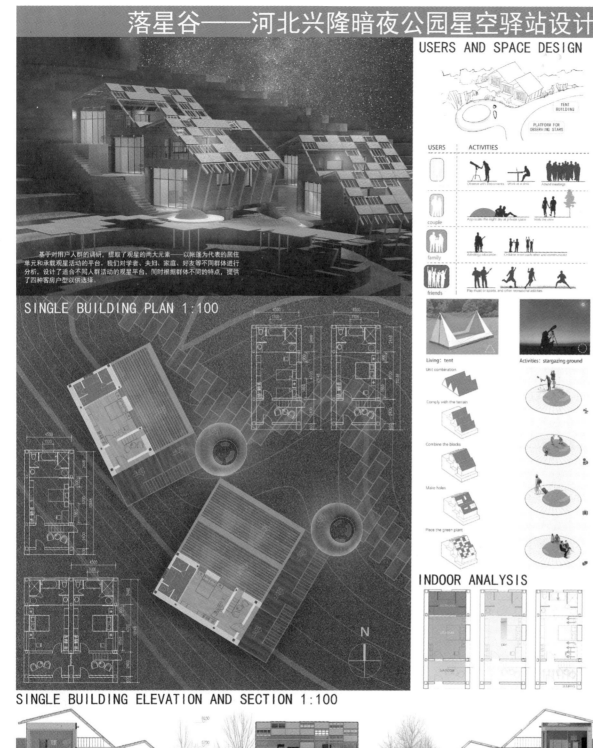

基于对用户人群的调研，提取了观星的两大元素——以帐篷为代表的居住单元和承载观星活动的平台。我们对学者、夫妇、家庭、好友等不同群体进行分析，设计了适合不同人群活动的观星平台，同时根据群体不同的特点，提供了四种客房户型以供选择。

THE VALLEY OF FALLING STARS

04

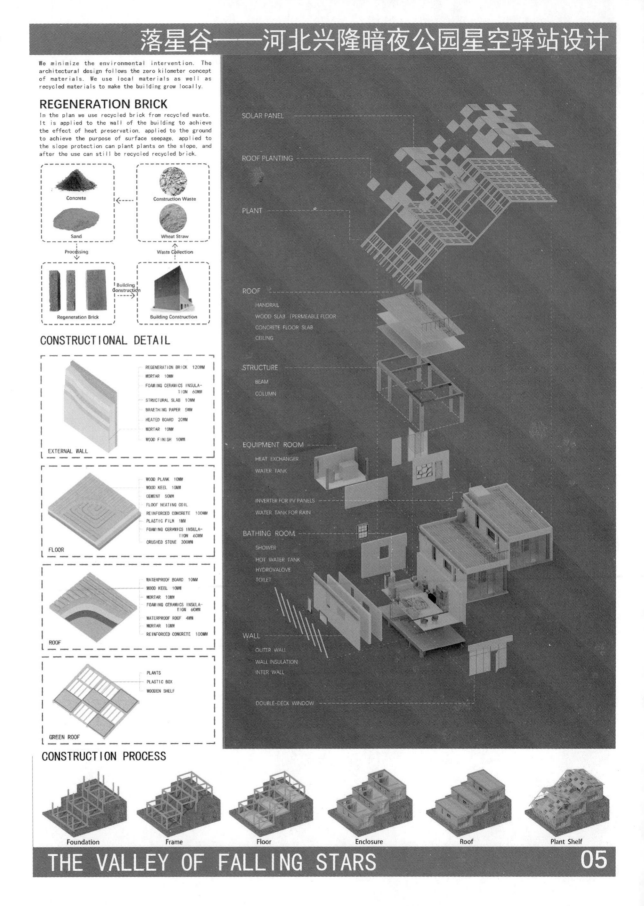

落星谷——河北兴隆暗夜公园星空驿站设计

EQUIPMENT OPERATION

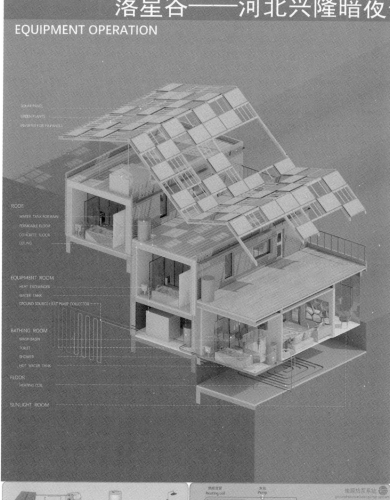

The building adopts a combination of active and passive methods to achieve the building's energy conservation goals.

PASSIVE TECHNOLOGY: Sunshine rooms are set in the south of the building to ensure indoor comfort. The planting roof facing west solves the problem of sun exposure. Large windows are opened in the southeast of the building, while small windows are opened in the northwest to ensure the natural ventilation of the building in summer and isolate the northwest wind in winter.

ACTIVE TECHNOLOGY: Through the combined operation of solar hot water system, ground source heat pump system, rainwater recycling system, photovoltaic system and biogas system to ensure the self-supply and recycling of building energy.

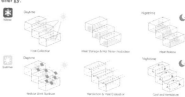

PRINCIPLE OF SUNLIGHT ROOM

Winter
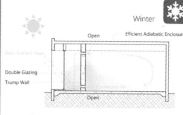
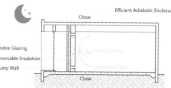

Summer
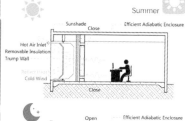
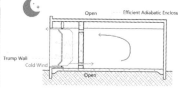

RAINWATER RECYCLING

The surface of the terrace is covered with permeable wooden boards. The rainwater goes down through the wooden frame and leaks down to the gutter.

Rainwater discharge through the parapet into the gutter, and is collected into the rainwater tank for secondary use by siphonic rainwater inlet.

Rainwater is collected into water tanks to irrigate plants, or into gutters for recycling.

THE VALLEY OF FALLING STARS — 06

综合奖·优秀奖
General Prize Awarded · Honorable Mention Prize

注 册 号：6199
项目名称：兴隆暗夜公园星空驿站 星语心"院"（兴隆）
The Starry Sky Station of XingLong Dark Night Park Dream Courtyard, Talking with Star, Wishing with Heart（Xinglong）

作　　者：王润中、骆祉安、周林龙、顾逸莹、王泽慧、熊亮亮、汪文正、贾星禄、张伟康、张逸飞、张海烨

参赛单位：上海济光职业技术学院

指导老师：李书谊、张宏儒、张 颖、宋雯珺、乔正珺、柏梦杰、俞 波、缪嘉晨、陈仙鸿

PERSPECTIVE DRAWING OF BIRD'S-EYE VIEW

星语心"院"
兴隆暗夜公园星空驿站　2019 台达杯国际太阳能建筑设计竞赛

DREAM COURTYARD, TALKING WITH STAR, WISHING WITH HEART　CHAPTER 1 CONCEPT & SITE ANALYSIS

■ DESIGN STATEMENT

本星空驿站项目位于河北兴隆暗夜公园内，设有酒店大堂和餐饮、会议中心、酒店客房（28间客房），总建筑面积为2455平方米。项目充分发挥山地建筑特色，依山而建。客房组团布局采取中国传统民居"合院"理念，体现"开放、共享、和睦"之美好愿景。采用光伏一体化建筑设计，实现"零碳建筑"理念。项目设计着手私密性、开放性的创意科普观星空间。简而言之，生态建筑，同自然共处；合院空间，与他人交流；星空平台，和自己对话。

The Starry Sky Station project is located at the Dark Night Park of XingLong County, Hebei. The project includes hotel lobby and catering, conference center, guest rooms (28 rooms), with total area of 2455 square meters. Following the mountain shape, the project gives full play to the characteristics of mountain architecture. The layout of guest rooms adopts Chinese traditional residence "courtyard" concept, which embodies the beautiful vision of "Openness, Sharing and Harmony". Integrated building design with photovoltaic achieves the "zero carbon building" concept. The project designs numbers of private and open creative Star-study space, in order to popularize scientific knowledge. Multi-dimensional space system: ecological architecture, co-existence with nature; Courtyard space, communicate with others; starry sky platform, talk to yourself.

■ DESIGN CONCEPT

Harmony
- With Nature
  - to make full use of the terrain
  - to protect vegetation and reduce excavation
  - to construct different landscapes according to different climatic conditions
- With People
  - to strengthen the communication between people
  - to fully consider the special needs of the disabled and the elderly

Context
- Architectural Form & Innovation
  - to create a new style of COURTYARD according to extraction of Typical Spatial Models in Northern China
- Traditional Materials & Innovation
  - to reproduce the performance of traditional materials with new materials
- Native Plants Species
  - to plant local tree species to reflect local landscape characteristics

Technology
- Zero-Carbon Building
- Sustainable Sites

■ FUNDAMENTAL CONDITIONS

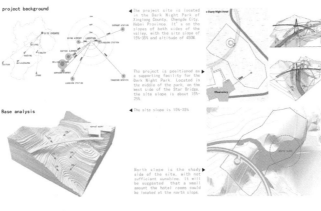

project background

The project site is located in the Dark Night Park of Xinglong County, Chengde City, Hebei Province. It's on the slopes of both sides of the valley, with the site slope of 15%-35% and altitude of 450M.

The project is positioned as a supporting facility for the Dark Night Park. Located in the middle of the park, on the west side of the Star Bridge, the site slope is about 15%-25%.

Base analysis

The site slope is 15%-25%.

North slope is the shady side of the site, with not sufficient sunshine. It will be suggested that a small amount of the hotel rooms could be located at the north slope.

MASTER PLAN

星语心"院"

兴隆暗夜公园星空驿站 2019 台达杯国际太阳能建筑设计竞赛

DREAM COURTYARD, TALKING WITH STAR, WISHING WITH HEART
CHAPTER 2 MASTER PLAN

TRAFFIC ANALYSIS SIGHT VIEW ANALYSIS

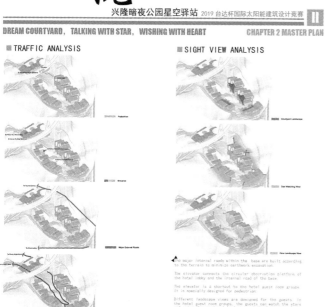

TECHNICAL AND ECONOMICAL INDEX

Item		Numerical Value	
Land Area		7952 m²	
Total Construction Area		2455 m²	
Hotel Lobby & Restaurant	930 m²	Hotel Lobby and Services Rooms: 295 m²	
		Restaurant, Kitchen, Lounge: 310 m²	
		Offices and Logistics: 325 m²	
Conference Center	285 m²	Meeting Room (2 rooms): 110 m²	
		Conference Center Lobby: 100 m²	
		Offices: 75 m²	
Hotel Guest Room Groups (4 groups)	1240 m²	Hotel Guest Room Groups: 1200 m² (4 groups in total, 300 m² for each group. Each group includes 7 units, 28 units in total.)	Hotel Guest Room Group A (3 Groups): 900 m² (Each group includes 7 units, 43 m² for each unit)
			Hotel Guest Room Group B (1 Groups): 300 m² (includes 7 units, 43 m² for each unit)
		Hotel Room Service: 40 m²	
Building Density		23%	
FAR Floor Area Ratio		0.31	
Green Space Ratio		46%	
Figure Coefficient		Hotel Lobby & Restaurant: 0.39	
		Conference Center: 0.35	
		Hotel Guest Room Groups: 0.38	

SITE WIND ENVIRONMENT ANALYSIS

1. Wind environment analysis model

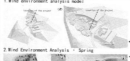

2. Wind Environment Analysis · Spring

3. Summer and autumn conditions

4. Winter conditions

5. Wind environment summary

RENEWABLE ENERGY: SOLAR ENERGY
Site Solar Radiation

1. Renewable Energy: Solar Water Heater

2. Renewable Energy: Photovoltaic

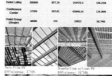
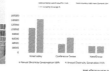

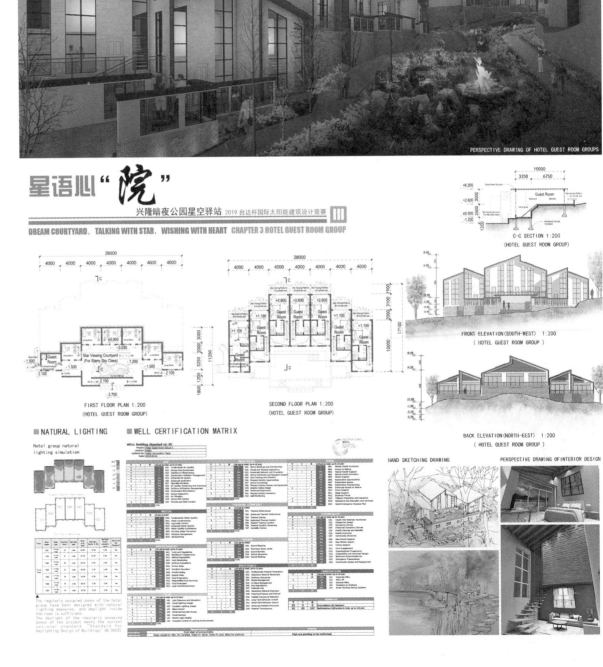

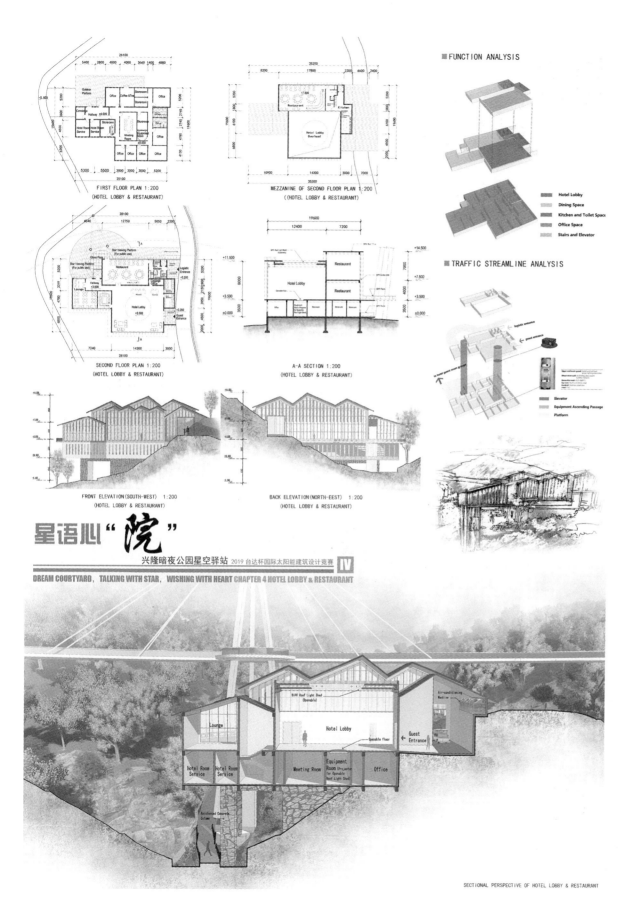

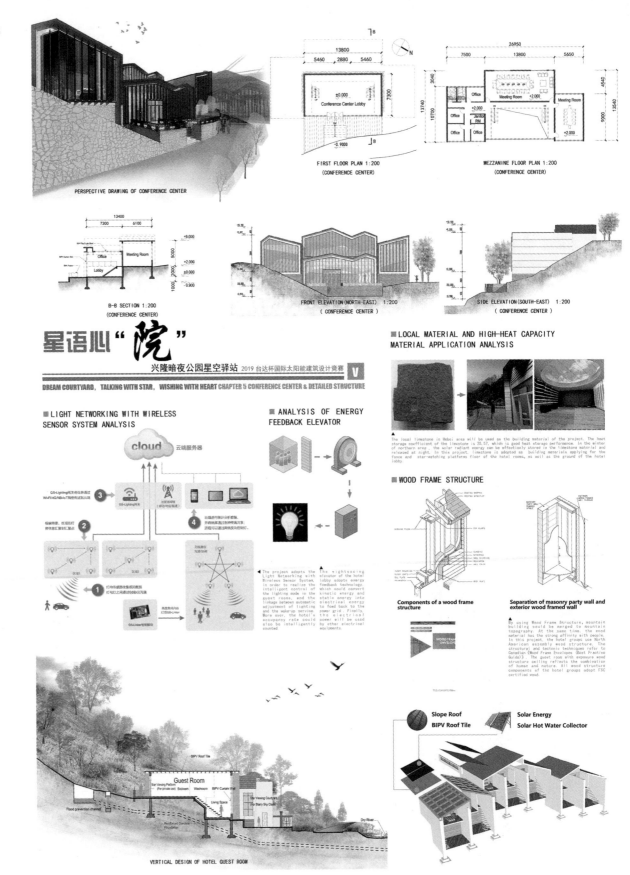

BUILDING INTEGRATED PHOTOVOLTAICS DESIGN

No.	ID Color	Building Component	Pattern	Product Dimensions	Building Name	Orientation	Quantity	Total Area (m²)	Area1 (m²)	Area2 (m²)
1	1	BIPV Curtain Wall		Area1: Standard Component: L1200*600*D7.0mm Area2: Nonstandard Component: Varied (BIPV Glass)	Hotel Guest Room Group A	South-East	109	79	69.84	9.16
					Hotel Guest Room Group B	South-West	109	79	69.84	9.16
				Area1: Standard Component: L1200*1800*D16.7mm Area2: Nonstandard Component: Varied (BIPV Glass)	Hotel Lobby & Restaurant	South/South-West	132	283.35	216	67.35
					Conference Center	South-West	71	152.3	101.5	50.8
2	2	BIPV Roof Light Shed		Area1: Standard Component: L1200*1800*D16.7mm Area2: Nonstandard Component: Varied (BIPV Glass)	Hotel Lobby & Restaurant	South-West	29	62.16	34.56	27.6
				Standard Component: L1200*1800*D16.7mm (BIPV Glass)	Conference Center	South-West	42	90.85	/	/
3	3	BIPV Roof		Standard Component: L1250*420*D4.2mm (BIPV Roof Tile)	Hotel Guest Room Group A	South-East	126	66	/	/
					Hotel Guest Room Group B	South-West	126	66	/	/
					Hotel Lobby & Restaurant	South/South-West	952	500	/	/
					Conference Center	South-West	263	138	/	/
4	4	BIPV Fence		Standard Component: L1200*600*D7.0mm (Manufacturing process refer to Detail Drawing) (BIPV Glass)	Hotel Lobby & Restaurant	South/South-West	44	31.68	/	/
					Conference Center	South-West	22	15.84	/	/
				Standard Component: L1200*600*D7.0mm (BIPV Glass)	Hotel Guest Room Group A	South-East/South-West	14	10	/	/
					Hotel Guest Room Group B	South-West	14	10	/	/
5	5	BIPV Window Grille (At Hotel Group Courtyard Wall)		Standard Component: L1200*600*D7.0mm (Manufacturing process refer to Detail Drawing) (BIPV Glass)	Hotel Guest Room Group A	South-East	28	20	/	/
					Hotel Guest Room Group B	South-West	28	20	/	/
6	6	Solar Hot Water Collector		Standard Component: L2000*1000*D80mm (Solar Hot Water Collector)	Hotel Guest Room Group A	South-East	21	42	/	/
					Hotel Guest Room Group B	South-West	21	42	/	/
					Hotel Lobby & Restaurant	South-East	5	10	/	/
					Conference Center	South-East	5	10	/	/

3D DECOMPOSITION DRAWING OF BIPV WINDOW GRILLE

ENLARGED FLOOR PLAN FOR GUEST ROOM

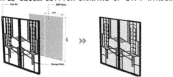
FIRST FLOOR PLAN
(NORMAL HOTEL GUEST ROOM)

SECOND FLOOR PLAN
(NORMAL HOTEL GUEST ROOM)

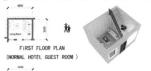
FIRST FLOOR PLAN
(HOTEL GUEST ROOM FOR DISABLED PEOPLE OR SENIOR PEOPLE)

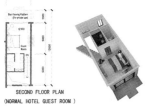
SECOND FLOOR PLAN
(HOTEL GUEST ROOM FOR DISABLED PEOPLE OR SENIOR PEOPLE)

星语心 "院"

兴隆暗夜公园星空驿站　2019 台达杯国际太阳能建筑设计竞赛 VI

DREAM COURTYARD, TALKING WITH STAR, WISHING WITH HEART CHAPTER 6 TECHNOLOGY & STAR VIEWING

ANALYSIS OF THE VERTICAL ADJUSTABLE SUNSHADING BOARD

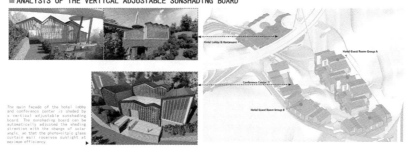

The main facade of the hotel lobby and conference center is shaded by a vertical adjustable sunshading board. The sunshading board can be automatically adjusted the shading direction with the change of solar angle, so that the photovoltaic glass curtain wall receives sunlight at maximum efficiency.

STAR-VIEWING ANALYSIS

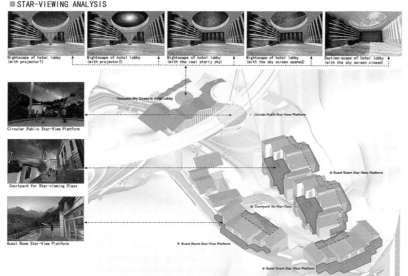

综合奖·优秀奖
General Prize Awarded · Honorable Mention Prize

注 册 号：6382
项目名称：星光·叠落（兴隆）
　　　　　Star-Light-Sprinkle
　　　　　（Xinglong）
作　　者：马金辉、林 霞、张传群
参赛单位：重庆大学
指导老师：周铁军、张海滨、唐鸣放

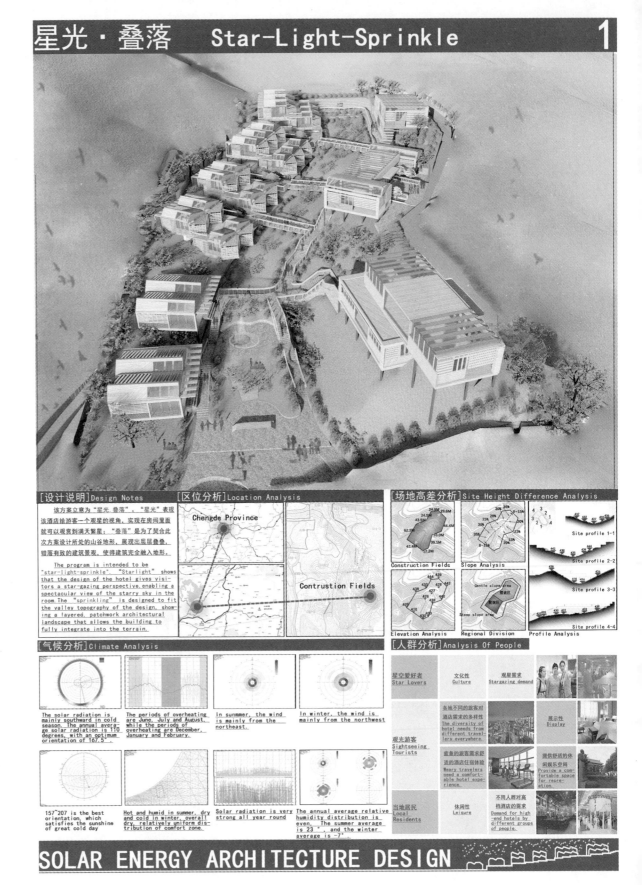

星光·叠落 Star-Light-Sprinkle

1. Single room
2. Double room
3. The dirt room
4. Laundry room
5. Storage room
6. General manager's office
7. Accounting office
8. Lobby bar
9. The restaurant entrance
10. The restaurant
11. Rooms
12. Work station

[一层平面图] First Plan 1:300

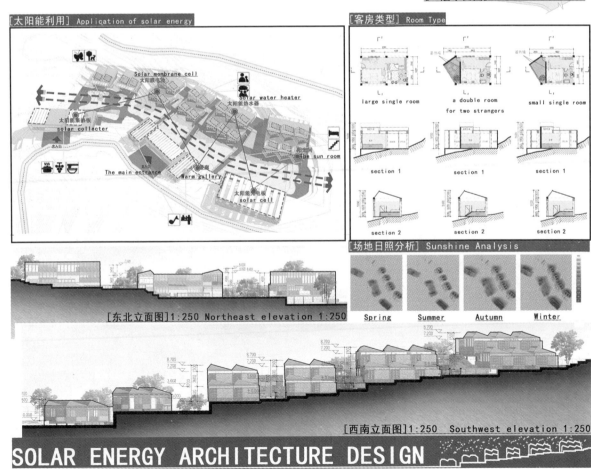

[太阳能利用] Application of solar energy

[客房类型] Room Type

[场地日照分析] Sunshine Analysis
Spring / Summer / Autumn / Winter

[东北立面图] 1:250 Northeast elevation 1:250

[西南立面图] 1:250 Southwest elevation 1:250

SOLAR ENERGY ARCHITECTURE DESIGN

星光·叠落 Star-Light-Sprinkle   4

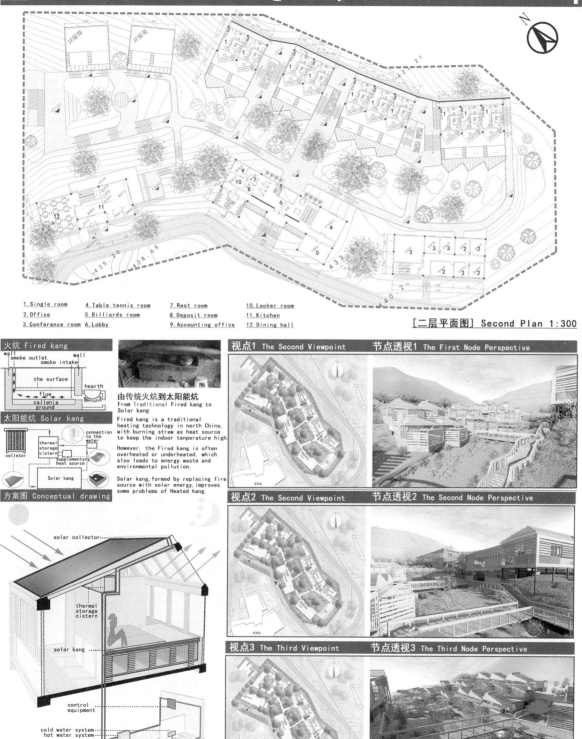

[二层平面图] Second Plan 1:300

1. Single room
2. Office
3. Conference room
4. Table tennis room
5. Billiards room
6. Lobby
7. Rest room
8. Deposit room
9. Accounting office
10. Locker room
11. Kitchen
12. Dining hall

由传统火炕到太阳能炕
From Traditional Fired kang to Solar kang

Fired kang is a traditional heating technology in north China, with burning straw as heat source to keep the indoor temperature high.

However, the Fired kang is often overheated or underheated, which also leads to energy waste and environmental pollution.

Solar kang, formed by replacing fire source with solar energy, improves some problems of Heated kang.

The room is heated by radiation and convention, and solar energy is used as the heat source to heat the floor through a coil laid on the floor.

视点1 The Second Viewpoint — 节点透视1 The First Node Perspective
视点2 The Second Viewpoint — 节点透视2 The Second Node Perspective
视点3 The Third Viewpoint — 节点透视3 The Third Node Perspective

SOLAR ENERGY ARCHITECTURE DESIGN

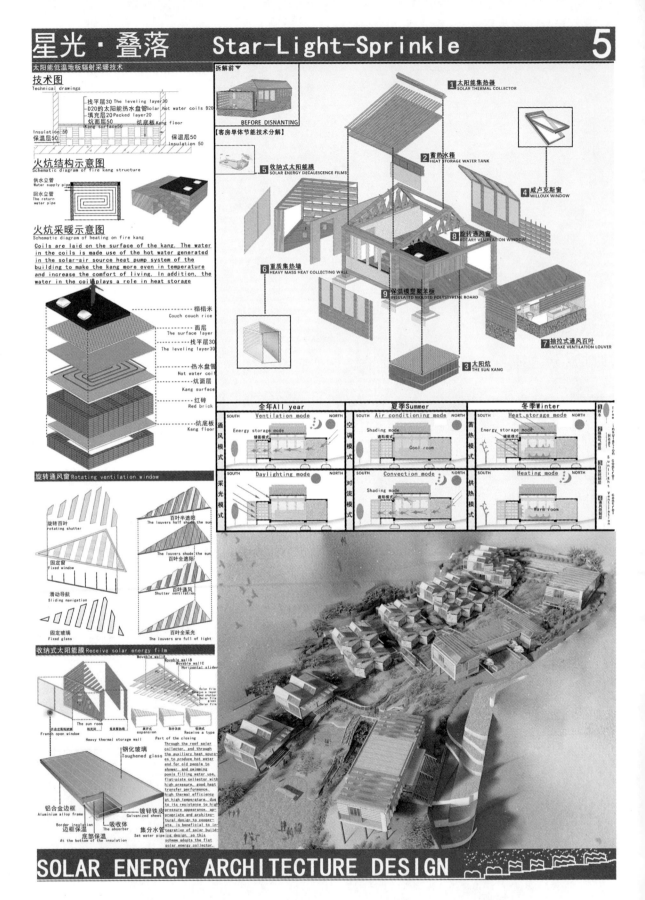

星光·叠落 Star-Light-Sprinkle

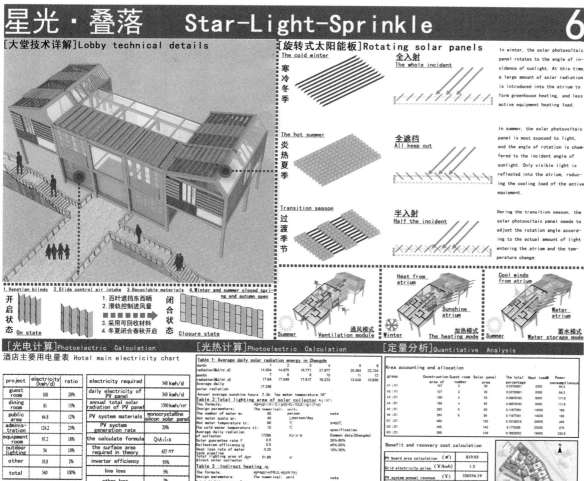

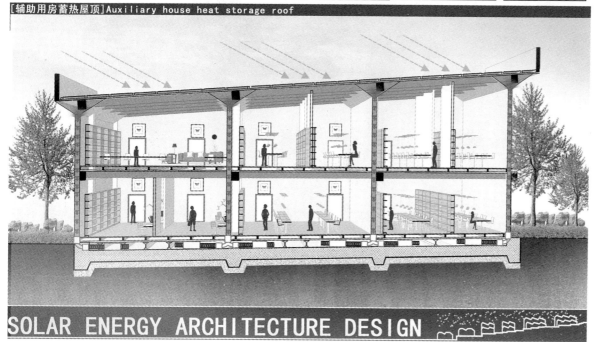

SOLAR ENERGY ARCHITECTURE DESIGN

综合奖·优秀奖
General Prize Awarded·
Honorable Mention Prize

注　册　号：6408
项目名称：一期一会（兴隆）
　　　　　The Very Moment (Xinglong)
作　　者：王博晓、张　颖、雷　坤
参赛单位：南京工业大学
指导老师：张伟郁、杨亦陵

DESIGN DESCRIPTION

场地位于山谷地带，南北狭长，东西陡峭，形成三面围合的空间。我们利用地势，将场地中间设计成景观廊道，置入亭子和步道来增加空间的层次感，达到步移景异的效果。通行和观赏空间的交错设计，使人们在穿梭过程中发生"期待——停留——期待——停留"的心理变化，也就是"一期一会"的概念。

建筑兼顾主、被动式的太阳能技术。依据当地气候特点进行建筑布局，并采用了生态雨洪，种植屋面等其他绿色技术。

场地内设计了不同层次的观星空间——集中广场、室外平台、私人庭院，充分满足观星爱好者的需求。

The site is located in the valley, long and narrow from north to south, steep from east to west, forming a enclosed space on three sides. We made use of the topography to design a landscape corridor in the middle of the site, and put pavilions and footpaths to increase the level of the space, so as to achieve the effect of different sceneries. The staggered design of pass-through and ornamental space makes people experience the psychological change of "expecting-staying-expecting-staying" during the shuttle process, which is also the concept of "The Very Moment".

The building combines active and passive solar energy technologies. According to the local climate characteristics of the building layout, and the use of ecological rain, planting roof and other green technology.

Stargazing space of different levels is designed in the site -- centralized square, outdoor platform and private courtyard, fully meeting the needs of stargazing fans.

一期·一会 The Very Moment

1 OVERAL IMPRESSION

BACKGROUND INFORMATION

LOCATION

Xinglong

The project is located in the Dark-Sky Park of Anyingzhai Village, Liudaohe Town, Xinglong County, Chengde City, Hebei Province, with a latitude of 40°41′ north and a longitude of 117°25′. The Dark-Sky Park is on the slopes on both sides of the valley, with an altitude of about 450 meters and a slope of 15%-35%. Surrounding the site are commercial streets, low-rise houses, apartments, and cultural and recreational facilities such as ski resorts and sports parks.

REGIONAL FEATURES

Enclosing Yard | Flush Gable Roof

Stone Pit | Timber

SITE

Location — The site is in the south east of the Park, close to central area and partly beneath the Starry Bridge.

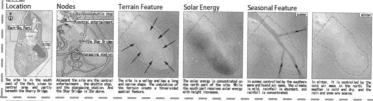
Nodes — Adjacent the site are the central entertainment, the shuttle stop, and the stargazing station. And the Star Bridge is 21m above.

Terrain Feature — The site is a valley and has a long and narrow shape. The undulation of the terrain creates a three-sided spatial feature.

Solar Energy — The solar energy is concentrated on the north part of the site. While the south part receives solar energy with height increases.

Seasonal Feature — In summer, controlled by the southern warm and humid air mass, the climate is mild, rainfall is abundant, and rainfall is concentrated. In winter, it is controlled by the cold air mass in the north. The weather is cold and dry, and the rain and snow are scarce.

CLIMATE ANALYSIS

Optimal Orientation | South Solar Radiation | Winter Solstice Day
Thermal Comfort | Wind Frequency | Weekly Temperature

CONCEPT
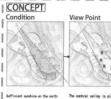 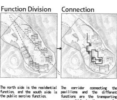
Condition | View Point | Function Division | Connection | Respondance | Pleasure

Sufficient sunshine on the north / Summer rain zone / Insufficient light on the south — The central valley is placed into an aerial view pavilion as a view corridor, which is the place for visitors to stop and interact with other visitors, the scenery or the starry night sky. — The north side is the residential function, and the south side is the public service function. — The corridor connecting the pavilions and the different functions are the transporting space, which is the space for expectations. — The arch of the portal is respond to the shape of the star bridge above the site. — The interlaced combination of traffic space and viewing space increase the pleasure for the guests inside.

一期·一会 The Very Moment

2 INTERGRATED LAYOUT

TRAFFIC ANALYSIS

outside | inside | fierfighting
— tourist / — staff

FUNCTION DICVISION
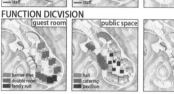
guest room | public space | rear service
barrier-free / double room / family suit | hall / catering / pavilion

STARGAZING SITES

most public | public | private

SITE PLAN 1:500

TECHNICAL AND ECONOMIC INDEX

Economic and Technical Norms			
Area of Base	8971.23 m²	Greening Rate	47.3%
Area of Structure	2482.36 m²	Plot Rate	0.28
Area of Site	1947.5 m²		

	Function	Number	Area	Gross Area
Guest Room	Standard Room	18	43.75 m²	787.5 m²
	Family Suite	3	76.38 m²	229.14 m²
	Accessible Room	4	43.75 m²	175.0 m²
	Room Service	2	30.25 m²	60.5 m²
Common	Hotel Lobby	1	133.5 m²	314.2 m²
	Duty Room	1	14.0 m²	
	Public Washroom	1	22.5 m²	
	Activities Hall	1	144.20 m²	
Catering	Kitchen	1	89.7 m²	399.3 m²
	Dining Room	1	146.8 m²	
	Washroom	1	30 m²	
	Bar	1	132.8 m²	
Administration	Administration office	3	54.3 m²	176.5 m²
	Meeting Room	2	60.2 m²	
	Joint Office	1	40.7 m²	
Rear Service	Storeroom	1	23.5 m²	

AERIAL VIEW

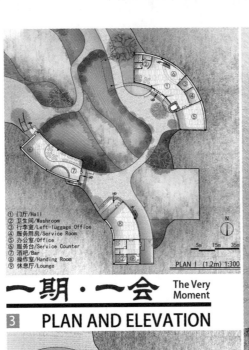

① 门厅/Hall
② 卫生间/Washroom
③ 行李室/Left-luggage Office
④ 服务用房/Service Room
⑤ 办公室/Office
⑥ 服务台/Service Counter
⑦ 酒吧/Bar
⑧ 操作室/Handing Room
⑨ 休息厅/Lounge

PLAN I (1.2m) 1:300

一期·一会 The Very Moment

3 PLAN AND ELEVATION

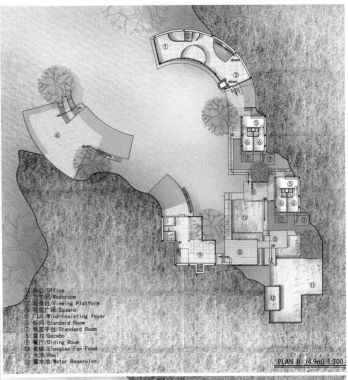

① 办公/Office
② 卫生间/Washroom
③ 观景台/Viewing Platform
④ 观星广场/Square
⑤ 门斗/Wind-resisting foyer
⑥ 标间/Standard Room
⑦ 观星平台/Standard Room
⑧ 露台/Gazebo
⑨ 餐厅/Dining Room
⑩ 食梯/Elevator for Food
⑪ 水池/Pool
⑫ 蓄水池/Water Reservior

PLAN II (4.9m) 1:300

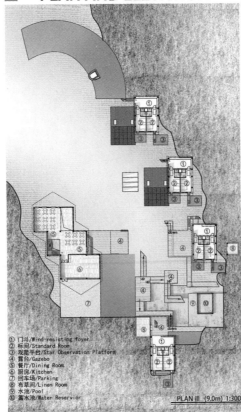

① 门斗/Wind-resisting foyer
② 标间/Standard Room
③ 观星平台/Star Observation Platform
④ 露台/Gazebo
⑤ 餐厅/Dining Room
⑥ 厨房/Kitchen
⑦ 回车场/Parking
⑧ 布草间/Linen Room
⑨ 水池/Pool
⑩ 蓄水池/Water Reservoir

PLAN III (9.0m) 1:300

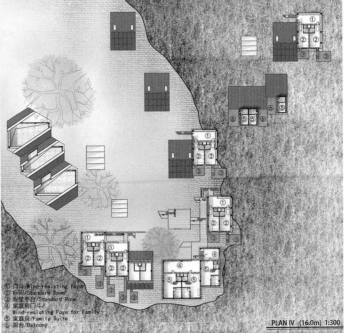

① 门斗/Wind-resisting foyer
② 标间/Standard Room
③ 观星平台/Standard Room
④ 家庭房门斗/Wind-resisting Foyer for Family
⑤ 家庭房/Family Suite
⑥ 阳台/Balcony

PLAN IV (16.0m) 1:300

NORTH ELEVATION 1:300

ENERGY SAVING TECHNOLOGY ANALYSIS

INTERGRATED LAYOUT | UNIT | INDIVIDUAL GUDSTROOM

LIGHTNING CONDITION | ZONE DIVISION | LIGHTNING CONDITION

Shadow Analysis in Certain Date | Eight Sections of Energy Supply | Optimization of Lightning / Natural Illuminance

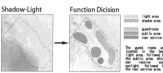

Shadow-Light → Function Dicision

The guest rooms are located in the best light area, followed by the public area, which can receive some sunlight, followed by the rear service area.

The site is divided into five sections. One pavilion with eight guest rooms is a unit in average. The energy resources of each room is connected to the pavilion through pipelines. The pavilions are connected through pipelines. Thereby achieving effective management of energy.

THERMAL ENVIRONMENT
Optimizaition of Wall Insulation

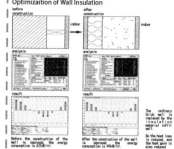

Before the construction of the wall is improved, the energy consumption is 67kW/m². After the construction of the wall is improved, the energy consumption is 49kW/m².

The ordinary brick wall is replaced by the insulation material infill wall.

So the heat loss is reduced, and the heat gain is also reduced.

SOLAR ENERGY | SOLAR ENERGY
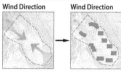
Solar Energy → Solar Panel Distribution

Solar Panel
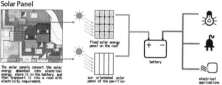

The solar panels are distributed in places where solar energy is the most concentrated during the year to get the most solar energy resources.

The solar panels convert the solar energy absorbed into electrical energy, store it in the battery, and then transport it into a room with electricity requirement.

WIND ENVIRONMENT | WIND ENVIRONMENT | WIND ENVIRONMENT
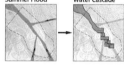
Wind Direction / Wind Direction

Adjustable Louvers

unfold to admit wind / fold to exclude wind

Wind Resisting Foyer
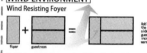
foyer + guestroom =

The building arrangement is folded in the northeast-southeast direction to admit the cooling summer breeze from the northwest and to exclude the cold winter wind from the southeast.

The adjustable louvers in the pavilions are unfold to admit the cooling summer breeze and fold to exclude the cold winter wind.
So that the microclimate will be more cozy.

Add a foyer in the north-east side of the two guestrooms to increase the warmth indoor.

Thick Stone Wall / Small Ventilation Window

WATER CONDITION | WATER CONDITION | RECYCLED MATERIAL
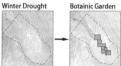
Summer Flood / Water Cascade

Water Reservoir
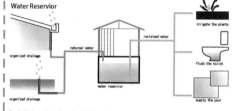
organized drainage / natural water → water reservoir → irrigate the plants / flush the toilet / supply the pool

Trees in the Site
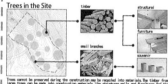
timber → structural / furniture
small branches → souvenir

In summer, controlled by the southern warm and humid air mass, the climate is mild, the rainfall is abundant and concentrated.
The pools in the valley gathers the rain and forms a scenery of falling water cascade.

Winter Drought / Botanic Garden

In winter, the weather is controlled by the cold air mass in the north. The atmosphere is cold and dry, and the rain and snow are scarce.
So, the waterlevel in the pools falls down beneath the plants under the water, presenting the scenery of botanic gardens.

Organized drainage and part of natural precipitation are piped into the reservoir below the pavilion. After sedimentation the water in the reservoir will be cleaned to be reclaimed water.
Then some of the reclaimed water are used to irrigate the plants and trees on the mountain slopes.
Some of the water are transported to the guestrooms or the rear-service in the restaurant, to flush the toilet or to wash the mops.
And the rest of the water are saved in the reservoir to supply the water in the pools in arid seasons.

Trees cannot be preserved during the construction may be recycled into materials. The timber from large trees can be made into construction materials like structures, walls and furnitures. And the smaller branches can be made into souvenirs for the guests.

一期·一会 The Very Moment

4 TECHNOLOGY SUMMARY

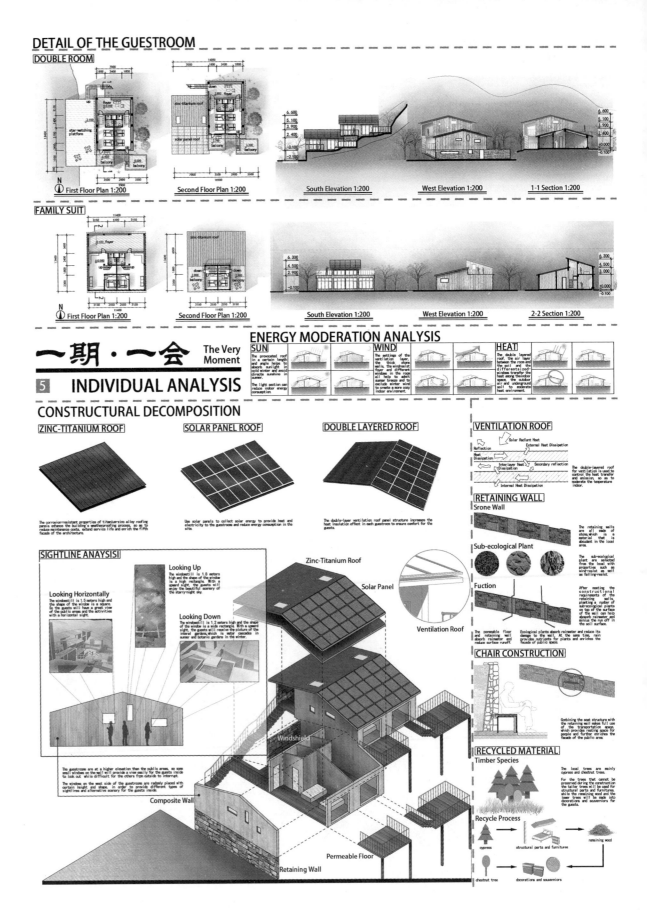

ECOLOGICAL RAINWATER PURIFICATION

DETAILED ECOLOGICAL RAINWATER PURIFICATION

一期・一会 The Very Moment

6 UNIT TECHNIC ANALYSIS

SEASONAL POOL

The pools that collect rainwater change with the seasons, and the dry season becomes a garden. The rainwater collected from the roofs, reservoirs, and permeable pavements during the flood season gathers here, flooding the bottom plants and becoming a water-storing landscape lake.

ECOLOGICAL TREE POOL

ECOLOGICAL RAINWATER PURIFICATION

PAVILLION DECOMPOSITION

Steel Structural Pavillion | Permeable Paving | Electrical Louver | Low-E Glass | Open | Partly Open | Closed

UNIT DECOMPOSITION

The roof collects rainfall and stores it in the reservoir after purification system.

Use the water pump to supply the collected rainfall into the guest room.

ENERGY SAVING STRUCTURE

LOW-E GLASS

Detailed Structure

Exegesis
① Film layer ⑤ Protective layer
② Glass layer ⑥ Glass layer
③ Antistatic film layer ⑦ Empty aluminum compartment
④ Polyester film layer ⑧ Fireproof sealant layer
  ⑨ Closed cavity

Application

Heat Loss 40% Indoor Radiation
Reflection 50% 10%
Solar Energy 100% Pass Through 65%

The application of low-e glasses in the thermal coating windows store the heat in the daytime and provide it to the guestrooms at night.

Through the glass layer at the upper and lower surfaces are respectively provided with glaze layer thermochromic coating. Adjusting the intensity of light changes the coating depth of color, so as to control the intensity of light through the glass, in order to control the indoor temperature by setting the layers.

The glass layer is provided in low-e toughened glass to increase the anti impact capability. Setting the protective film on glass under impact alleviates the impact force of glass by stretching the PVB diaphragm, thereby preventing the damage and rupture of glass.

PLANTING ROOF

Detailed Structure

Exegesis
① Vegetation layer
② Planting soil
③ Drainage collection system
④ Root-resistant waterproof layer
⑤ Ordinary waterproof layer
⑥ Building roof

Application
Pulic Star Gazing Ground

Main Hall Roof

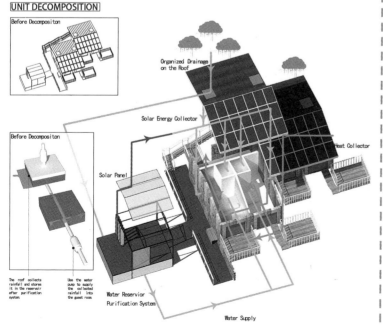

BAMBOO CURTAIN

Flexible shade without impact on viewing.

PLANTING WALL

Outdoor planting with wall decoration.

WHEN EARTH HUGS COURTYARD —— DESIGN OF HEBEI XINGLONG NIGHT PARK STAR STATION

综合奖·优秀奖
General Prize Awarded · Honorable Mention Prize

注　册　号：6510
项目名称：壤院（兴隆）
　　　　　　When Earth Hugs Courtyard (Xinglong)
作　　　者：林瑾如、李琴、徐寅莹、
　　　　　　洪啸林、李孟圆
参赛单位：西南交通大学
指导老师：李百毅

SITE LOCATION ANALYSIS
LOCATION
Location　　Site topography　　Land area

THE DARK NIGHT PARK PROJECT
The project land is located in the dark night park of Anyingzhai village, Liudaohe town, Xinglong county, Chengde city, Hebei province, 30 kilometers away from xinglong county. The dark night park is located on the slopes on both sides of the valley, about 450 meters above sea level, with a slope of 15%-35%.

Climate and topographic analysis

CLIMATE AND TOPOGRAPHIC ANALYSIS
Psychrometric　Temperature　Slope Analysis　Elevation Analysis
Orientation　Wind Frequency　Wind Temperatures　Relative Humidity

WIND DIRECTION
Northwest monsoon　Valley wind　Valley wind
winter　daytime　daytime
Southeast monsoon　Peak wind　Peak wind
summer　night　night

DESIGN CONCEPT
Scientists — The scientists' private activities at the star station are rest, stargazing and work
Tourists — The activities of tourists are recreation, stargazing and group building. They all need private space and viewing space

Their group activities are discussions and meetings

REQUIREMENT ANALYSIS

DESIGN CONCEPT

DESIGN NOTES
传统覆土建筑、四合院及庭院是北方冬季保温蓄热和夏季通风降温最为重要的建筑形式。
本方案从传统生态建筑策略"覆土"、"四合院"、"院落"出发，利用太阳能，解决北方山地环境带来的采暖、通风、采光等问题。利用土壤蓄热、特朗伯墙及阳光房等被动式策略解决采暖，利用合院、天井与建筑平面和体型设计引导气流，促进通风，解决山地建筑通风与采光问题。主动设计上，采用太阳能光伏光热系统，集热百叶等，为使用人群提供适宜的物理环境，创造生态、节能、宜人的居住与公共空间。

Traditional earth-covered buildings, siheyuan and courtyards are the most important building forms in winter heat preservation and summer ventilation. This scheme is based on the traditional ecological building strategy of "covering soil" and "siheyuan", "courtyard". Solar energy is used to solve the problems of heating, ventilation and lighting brought about by the northern mountainous environment. Passive strategies such as soil thermal storage, Trumber wall and sunroom are used to solve heating problems. Siheyuan, courtyard and building plane and shape design are used to guide air flow, promote ventilation, solve mountain building ventilation and lighting problems. In active design, solar photovoltaic photovoltaic system, collector blinds and so on are used to provide suitable physical environment for users and create ecological, energy-saving, pleasant living and public space.

ECONOMIC AND TECHNICAL INDEXES
Site Area: 7870 m²　　Green Area: 4170 m²　　Building Foot Print: 2550 m²
Building Area: 2720 m²　　Green Rate: 53%　　FAR: 0.34

 "Caves covered with soil + Siheyuan"　 "Caves covered with soil + Back yard"　 "Caves covered with soil + atrium"　 "Caves covered with soil + side yard"

We can take advantage of each other's ecological and energy saving advantages, and also complement each other. For example, heyuan can solve the problem of ventilation and lighting in caves.　　Adding a courtyard behind the cave allows for ventilation and light.　　A atrium in a public building will improve lighting and promote ventilation.　　Adding a courtyard beside the cave allows for ventilation and light.

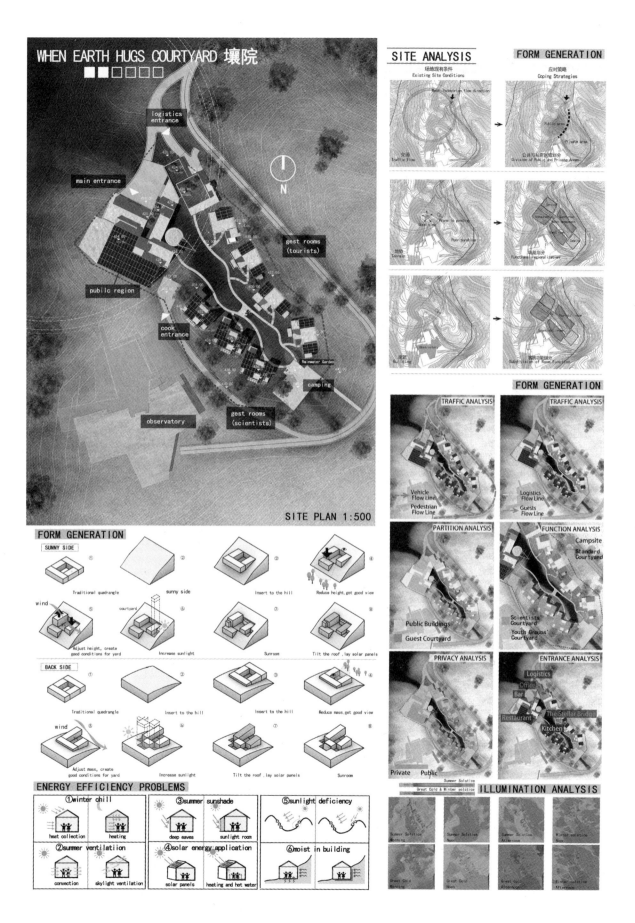

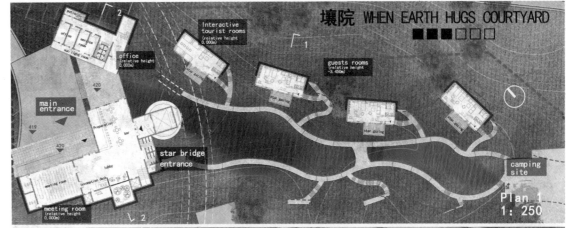
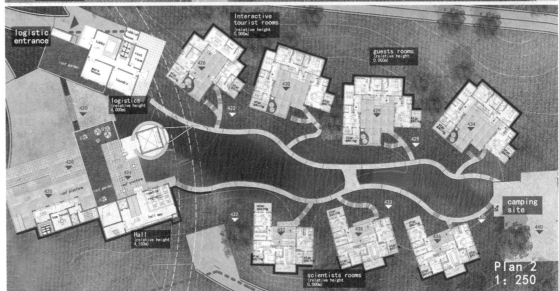

WHEN EARTH HUGS COURTYARD 壤院

South Hotel

North Hotel

Public House

UNIT ANALYSIS

Standard Courtyard Plan   1:150

Considering the number of tourists who are friends, family, couples, single people and so on. A standard courtyard consists of a single room, two double rooms and a family room.

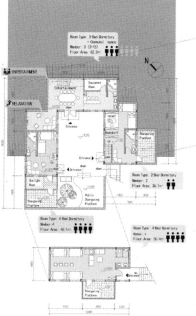

Youth Groups' Courtyard Plan   1:150

The rooms for young travelers are equipped with Communal living room in the northern apartment type, which can be used by all passengers in a courtyard.

Attic Floor Plan 1:150

Scientists' Courtyard Plan   1:150

The Scientists' Courtyard has three single rooms for scientists and scholars to live in. Public discussion rooms are set up in each courtyard to facilitate academic communications.

Accommodation Needs Analysis

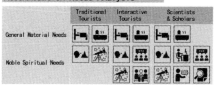

STARGAZING PLATFORM

scientists and scholars courtyard

youth groups' courtyard plan

stardard countyard plan

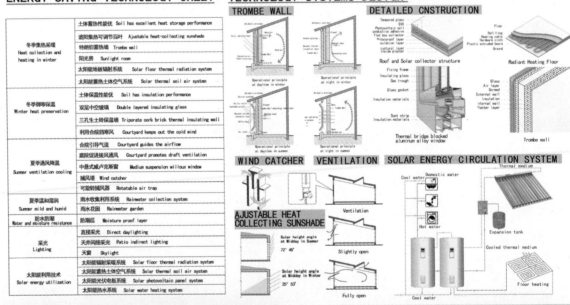

northeast elevation 1:400

southeast elevation 1:400

WHEN EARTH HUGS COURTYARD 壤院

SUNNY SIDE
WINTER ROOM STRATEGY

OVERVIEW OF TECHNIQUES

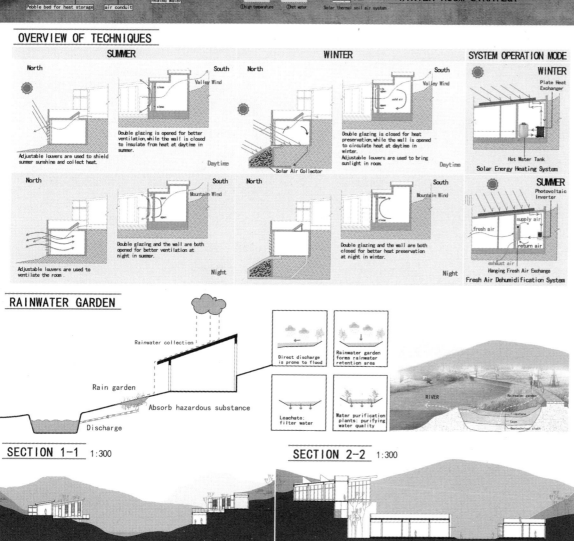

SECTION 1-1 1:300

SECTION 2-2 1:300

有效作品参赛团队名单
Name List of all Participants Submitting Effective Works

注册号	作者	单位名称	指导人	单位名称
5732	陈启东、雷博云、陈华雯、李凯、高锟硕	西安建筑科技大学	李涛	西安建筑科技大学
5741	李兆平、庚旭、冯向荣	北京交通大学	石克辉	北京交通大学
5745	寇朦心、康夏田、王幸珂、马艳、张茜	西安科技大学	孙倩倩	西安科技大学
5747	何雨微、霍文婕、徐紫莹、李子儒、张颖怡、刘瑞雪	西安科技大学	孙倩倩	西安科技大学
5767	杨娜、李伟、马昌华、杨嘉瑶、黄锐聪、王兰	西安科技大学	孙倩倩	西安科技大学
5770	杜天慧、刘泼、王海燕、张宁、尹雪、秦世位	华北理工大学	檀文迪	华北理工大学
5775	张昊、崔宁、王文辉	石家庄铁道大学	高力强	石家庄铁道大学
5790	胡庆强、刘思遥	重庆大学	周铁军、张海滨	重庆大学
5798	李俊杰、邹鸣鸣、袁佩桦、陈文华、王伟	东南大学、汉能移动能源集团	杨维菊	东南大学
5799	宋韵资、梁志伟、郭子杭、吴艾霖、曹原	中国矿业大学（北京）	李晓丹	中国矿业大学（北京）
5800	赵令宽、朱体强	河北建筑工程学院	王金奎	河北建筑工程学院
5807	黄承佳、江逸妍、汤雪儿、杜森、王文政、杨浩、翁杨华、刘璟彦	华南理工大学		
5809	孙敏、侍星宇、李佳乐	南京工业大学	陈军、林杰文	南京工业大学
5817	樊劲松、王珏、于志鹏	东北石油大学	李静薇	东北石油大学
5822	王子嘉、谢颂、苏晓婉	北京交通大学	杜晓辉、胡映东	北京交通大学
5826	覃煜、奚旺、袁亦	南京工业大学	胡振宇	南京工业大学
5835	周柳齐、郑若尘、田雪飞	南京工业大学	陈军	南京工业大学
5876	易慧慧、塔娜、高旭宏	内蒙古科技大学	白胤	内蒙古科技大学
5887	彭瑞辰、郭依奇	西安建筑科技大学		
5900	宁尹盈、林煜芸、黎海虹	广州大学	李丽、夏大为	广州大学
5902	李祖超、叶泳仪、曹伟龙	广州大学	李丽、夏大为	广州大学

续表

注册号	作者	单位名称	指导人	单位名称
5905	黎晓楠、刘文聪、陈璧璇、陈映红	广州大学、华南理工大学广州学院	李丽、裴刚	广州大学
5912	王睿欣、吴晓鹏	广州大学	李丽、夏大为	广州大学
5920	高海伦、方鑫磊、沈令婉、梅博涵、董兆、季缘	华南理工大学，浙江大学	王国光、王洁、浦欣成	华南理工大学、浙江大学
5925	徐小镇、池彦琳、林存钰	广州大学	李丽、夏大为	广州大学
5933	乔壬路	天津大学	刘彤彤	天津大学
5938	王杰汇、马振雷、姜子信、巫恋恋、杨心悦	天津大学	郭娟利	天津大学
5942	曹锦波、蒋群歆、徐敏、肖歆	广州大学、华南理工大学	李丽、周孝青	广州大学
5945	秦朗、田春来	重庆大学	周铁军、张海滨	重庆大学
5949	高伟杰、罗玥琪、黄方百、董学岭、赵国胜、崔瑶	华北理工大学	檀文迪	华北理工大学
5971	王佳佳、赵嘉健	石家庄铁道大学、中铁建安工程设计院西安分公司	高力强	石家庄铁道大学
6001	夏斯涵、刘颖杰、主曼婷、潘璐冉	安徽建筑大学	王薇	安徽建筑大学
6003	付彤雨、吕梦、崔向国	南京工业大学	罗靖	南京工业大学
6009	韩文颢、刘璐瑶	天津大学	贾巍杨、王小荣	天津大学
6016	周春妮、陈晗、程吉帆、陈胜蓝、曾港俊	宙思建筑设计（上海）事务所		
6017	谌思航、钟声、刘震、陈力欣	石家庄铁道大学	髙力强	石家庄铁道大学
6021	陈碧云、何周、武晓珊、杜康、秦超鹏	南阳理工学院	赵敬辛、高力强	南阳理工学院、石家庄铁道大学
6023	冯书雅、李晶晶、郭晓飞、王瑜鑫、赵妍、张文馨、赵信	石河子大学	王蒙、李靖	石河子大学
6024	刘如梅、杨飒爽、王洁妤、李瑾、蒋舒月	石河子大学	李洁	石河子大学
6025	钱程、王聿青、李尧、崔晓晨	西安建筑科技大学	靳亦冰	西安建筑科技大学
6031	张芷薇、刘璐、陈嘉雯	东北石油大学	徐晓丽、薛婷、李静薇	东北石油大学

续表

注册号	作者	单位名称	指导人	单位名称
6032	王博达、田佳、沈梦丹	华中科技大学、武汉理工大学	徐燊	华中科技大学
6034	韩依琪、李紫嫣、张文硕、赵雨晗、吴新汶、林卓敏	四川大学	王军、曾艺君	四川大学
6036	丁一靖、刘晨蕾	内蒙古科技大学	左云、贺晓燕	内蒙古科技大学
6041	张潇方、邱丛丛、陈家乐	武汉大学	黄凌江、李鹃	武汉大学
6044	程梦琪、洪叶、刘晶、范静哲、初楚、李佳宸、高铖	南昌大学	叶雨辰、郭兴国	南昌大学
6052	苏珊、王佳璇、叶芷茵	广州大学	李丽、万丰登、夏大为	广州大学
6053	马锡栋、江宏靖、陈可、张志豪	浙江理工大学	李茹冰、刘海强	浙江理工大学
6056	高源、刘雨田、梁向南、赵鑫、魏辛采	山东建筑大学	房涛、李晓东	山东建筑大学
6066	马家慧、吴晓嘉	武汉大学	黄凌江	武汉大学
6069	张莉	河南财经政法大学	吴孜淳	河南财经政法大学
6078	郦犊、孙钱森	浙江理工大学	毛万红	浙江理工大学
6084	孙浩、王晓晗、潘秀云、葛星、周晨、铁文艳	潍坊学院	韩升升	潍坊学院
6085	张福超、李升东、高洪、潘德强、梁志杰	兰州理工大学	赵丽峰、罗军瑞	兰州理工大学
6087	斯依提艾力·艾麦提、安通、张亚茹	新疆大学	王万江	新疆大学
6090	陈璐、高标、方心怡	南京工业大学	林杰文	南京工业大学
6093	李瑞娟、隋意、唐红宝、张市娇、冷柏畅、夏梦迪	潍坊学院	鞠洪磊、于兰兰	潍坊学院
6097	刘家韦华、张玉琪、郭嘉钰、胡紫雯	天津大学、河北工程大学	田芳、侯万钧	天津大学、河北工程大学
6100	王鑫琪、李可、魏迪,	华中科技大学		
6108	杨志祥、杨晓艳	天津大学	李伟、杨崴	天津大学
6111	杨裕雯、纪宗廷、张子安	南京工业大学	陈军、林杰文	南京工业大学
6112	郭佳奕、王钰杰	武汉大学	李鹃	武汉大学

续表

注册号	作者	单位名称	指导人	单位名称
6116	刘振生、左方圆、梅强	华中科技大学	徐燊、万谦	华中科技大学
6117	王丽萍、黄芷璇	武汉大学城市设计学院	兰兵	武汉大学城市设计学院
6121	王少杰、丁鬲、柳天元	沈阳建筑大学 丝路视觉科技股份有限公司	王常伟、汝军红、王飒	沈阳建筑大学
6122	谢瑞杭、黄圣翔、叶珍光、马金辉	重庆大学	周铁军、张海滨	重庆大学
6127	朱海明、冯杰、顾皓楠	沈阳建筑大学	王常伟、汝军红	沈阳建筑大学
6128	保娟娟、尹榆珺	天津大学	王楚尧、崔宇茉	天津大学
6129	胡宇、张夏楚	天津大学	朱丽	天津大学
6130	李巧、钟莹	华南理工大学	肖毅强、王静	华南理工大学
6132	王哲雯、周楚来、赵楠茜	东北石油大学	薛婷	东北石油大学
6133	蔺朗、江玉璇、杨瀚康、李昊珉	北京交通大学	杜晓辉、胡映东	北京交通大学
6136	康艺兰、袁月、梅振、石帅	上海大学	张维	上海大学
6144	葛葳、乔亚萍、袁振皓	山东建筑大学	纪伟东	山东建筑大学
6146	从琳、高宇轩、谢恩枫、赵旭丹、南文娟、张许山、王鲁拓、宋林晨	北方工业大学	马欣、赵春喜	北方工业大学
6152	闫宇航、王钰、王雪	东北石油大学	李静薇、徐小丽、薛婷	东北石油大学
6153	赵茂均、董鑫	重庆大学	周铁军、张海滨	重庆大学
6156	许锦灿、史小宇、吴佳晋、周浩	南京工业大学	张伟郁、杨亦陵	南京工业大学
6159	彭硕、刘震、吕紫薇、何小钰、陈力欣、王雨博、温政烨	石家庄铁道大学	髙力强、李佳琳	石家庄铁道大学
6167	杨淋琳、刘志平、刘雯雨、潘纪元、张雪媛、叶馨宇	中国矿业大学	杨灏	中国矿业大学
6173	赵谷橙、岳小超、齐梦晓、李志权、金宁园、王晨	河北工业大学	李有芳	河北工业大学
6175	徐杭、杨佳明、张凯珂、韩菁菁、李曦、吴艳	华侨大学	冉茂宇	华侨大学

续表

注册号	作者	单位名称	指导人	单位名称
6181	张杰、杨晓琳、杨鑫豪	沈阳建筑大学	王常伟、汝军红、张圆	沈阳建筑大学
6182	金淏文、段辉、李卉馨、张谡艺	大连理工大学	李国鹏	大连理工大学
6183	石丹、耿艺曼	重庆大学	周铁军、张海滨	重庆大学
6186	江海华、张晨、谢楠、宣姝颖	华中科技大学	徐燊	华中科技大学
6194	温政烨、王彬翰、何小钰、彭硕	石家庄铁道大学	高力强	石家庄铁道大学
6196	王强、李志信、马昕、温馨	华中科技大学	徐燊	华中科技大学
6198	王俞君、花晨、张玉珏、季新红、孙昊	盐城工学院	王进、荀琦	盐城工学院
6199	王润中、骆祉安、周林龙、顾逸莹、王泽慧、熊亮亮、汪文正、贾星禄、张伟康、张逸飞、张海烨	上海济光职业技术学院	李书谊、张宏儒、张颖、宋雯珺、乔正珺、柏梦杰、俞波、缪嘉晨、陈仙鸿	上海济光职业技术学院、上海市建筑科学研究院绿色建筑与低碳发展研究所、上海建科建筑设计院有限公司
6200	彭骆、易秋凡、王晨阳	华中科技大学	徐燊	华中科技大学
6205	赵一帆、寇俊涛、余南松、胡江博	南阳理工学院	赵敬辛、高力强	南阳理工学院、石家庄铁道大学
6216	陈秋实、王佳蕙、陈玉叶	华南理工大学	肖毅强、王静	华南理工大学
6225	王旭文、周元元、吴学俐、邵宗盖	大连大学	黄世岩	大连大学
6236	王庆朵、周枭、谭小静	山东建筑大学	陈兴涛	山东建筑大学
6238	陶萍、邵美璇、李鹏程	华中科技大学	杨毅	华中科技大学
6240	徐琪、李艳秋、吴芹彩、张柯强	南阳师范学院	王恒、高蕾	南阳师范学院
6242	陈灏、丁凯琪、王曦、侯荣婧、张静婷、邹济锴	中原工学院		
6249	张琮、王飞雪、张吉强、杨洋	天津大学	朱丽、孙勇	天津大学
6252	仲文洲、刘巧、隋明明、李心怡	东南大学、东京工业大学	张彤	东南大学
6253	Xia Wei（魏霞）、Kornel Mierzejewski、Dominik Nowicki、Radosław Ptaszyński、Jinzhong Wang（王金钟）、Moghazi Farrag、Peng Liu（刘鹏）	波兰波兹南理工大学		

续表

注册号	作者	单位名称	指导人	单位名称
6255	Xia Wei（魏霞）、Kornel Mierzejewski、Dominik Nowicki、Radosław Ptaszyński、Jinzhong Wang（王金钟）、Moghazi Farrag、Peng Liu（刘鹏）、Dominik Nowicki	波兰波兹南理工大学		
6256	黄现璁、杨韬、林广宇	上海大学	张维	上海大学
6258	金波、陈靖翎、郑俊超、张程博	浙江大学	金方	浙江大学
6263	朱珺、冯雪飞	天津大学	李伟	天津大学
6265	陈嘉蓉、计少敏、郑裕莹	广州大学	李丽	广州大学
6267	卞云菁、胡延根、敖圆圆	南京工业大学	陈军	南京工业大学
6269	顾植琴、李鹏娜、武桐、唐鑫	文华学院	余巍	文华学院
6280	柳旭、黄俐颖、高红霞、朱柯、刘娟娟	中国矿业大学、北京维拓时代建筑设计股份有限公司	李晓丹	中国矿业大学
6281	王俊棋、张沅杰、邓铱瑾、许萌洋	中央美术学院	苏勇	中央美术学院
6283	刘昌励	三川建筑事务所		
6285	李烨、卫冕、李伟	重庆大学	周铁军、张海滨	重庆大学
6289	康永基、王润丰、王子禛、郭雪婷	天津大学、河北工业大学	戴路、汪江华、舒平	天津大学、河北工业大学
6290	李彤、包嘉敏、陈成宇、易芳蓉、汤浩恒	华南理工大学	王静、肖毅强	华南理工大学
6292	丁富松、刘洁琼、何东明	南阳师范学院	高蕾	南阳师范学院
6293	张广远、侯宇晴、张星、王思琪	大连大学	王彦栋	大连大学
6296	巩昱澎、刘阳、李文斐、牛玥、孙腾飞、张翔宇、杨梦蝶	济南大学	王玲续、于江	济南大学
6304	黄旭	山东建筑大学	刘建军	山东建筑大学
6305	何小钰、彭硕、刘震、温政烨	石家庄铁道大学	高力强、李佳琳	石家庄铁道大学
6308	谢娜、丛欣宇、王萍	东北石油大学	李静薇	东北石油大学
6314	夏毓翎、赵雨、朱曦、鲍慧敏	南京工业大学	罗靖、杨亦陵	南京工业大学

续表

注册号	作者	单位名称	指导人	单位名称
6315	彭硕、陈力欣、吕紫薇、刘震	石家庄铁道大学	高力强、李佳琳	石家庄铁道大学
6320	刘孜业、胥婷婷、徐璨、李洋、刘洋	苏州科技大学	王依明、邱德华	苏州科技大学
6323	曹博、余欣珂、张琪曼、唐小全	中原工学院	孔莹博	中原工学院
6324	张帆、郭镒恺	山东建筑大学	陈兴涛	山东建筑大学
6327	柴宇豪、曲宝琪、闫帅、刘辰	山东建筑大学	江海涛，赵学义，魏琰琰	山东建筑大学
6334	吴子豪、石张睿、谢怡帆、胡峻语	苏州大学	赵秀玲	苏州大学
6340	黄嘉欣、王鹏举、殷娟、袁晓春	文华学院	余巍	文华学院
6343	谌穗、张晨、陈芷昱、龚胜泽	文华学院	余巍	文华学院
6350	金睿、吕窈瑶、谢正静、滕菲、胡庆玮	嘉兴学院	王中锋	嘉兴学院
6353	桑梦城、郑杜月、郑贵祥、朱婷	湖北工业大学	张辉	湖北工业大学
6355	倪平安、朱文瀚、宋克昌、吕城、巴赫蒂亚尔·多里坤	新疆大学	王万江	新疆大学
6360	武浩然、巩振华、吴昊、梁英伟	华南理工大学	肖毅强、王静	华南理工大学
6361	蔡晴雯、高丹、李金鸽、任飞宇	辽宁石油化工大学	潘剑	辽宁石油化工大学
6364	韦海燕、余沛颖、李娜	上海大学、上海美术学院	张维	上海美术学院
6374	贾兆元、李颖、刘佳雯、孔祥慧、何星辰、徐天阳、周畅、徐建立	北方工业大学	马欣、赵春喜	北方工业大学
6375	安然、王晴、管畅、张玮靓	天津大学、北京建筑大学	杨崴	天津大学
6376	芦俊聪、吴妆庄、颜婧、罗泽宇、梁景怡	广州大学	李丽、裴刚	广州大学
6382	马金辉、林霞、张传群	重庆大学	周铁军、张海滨、唐鸣放	重庆大学
6390	江静雯、纵天笑阳、历拔山、贺欢	西南交通大学	王俊	西南交通大学
6391	李雅、彤常远、何小钰、温政烨、彭硕、郭娇	石家庄铁道大学	高力强	石家庄铁道大学
6392	宋汪耀、姚歌、滑维杰	重庆大学	周铁军、张海滨	重庆大学

续表

注册号	作者	单位名称	指导人	单位名称
6404	杜静、段曰辉、李雨涵、黄沛沛	潍坊科技学院	李云云	潍坊科技学院
6405	周恒、李思亚、庞羽翔、杨晓蕾、王子平	安徽工业大学	黄志甲	安徽工业大学
6408	王博晓、张颖、雷坤	南京工业大学	张伟郁、杨亦陵	南京工业大学
6410	段炳好、钟秋阳	重庆大学	周铁军、张海滨	重庆大学
6412	白安民、杜袁昊、炳新宇	东北石油大学	李静薇、徐小丽、薛婷	东北石油大学
6419	温政烨、何小钰、李元、王雨博、王子安	石家庄铁道大学	高力强	石家庄铁道大学
6420	叶天爽、温腾飞、谢剑飞、田毓丰	苏州科技大学	邱德华	苏州科技大学
6421	何小钰、温政烨、刘震、王子安、李雅彤	石家庄铁道大学	高力强、李佳琳	河南大学建筑学院、石家庄铁道大学
6426	马睿、明玉、罗雨寒、张晓松	北方工业大学	于金鹭	北方工业大学
6431	苟新瑞、杨晨艺、陈志强、韩超	山东建筑大学	侯世荣	山东建筑大学
6432	黄兆旭	武汉科技大学		
6453	郭亚凝	西南交通大学	王蔚、王及宏	西南交通大学
6457	刘豪、刘复丹、隽永旭、孟浩	潍坊科技学院	崔晓、李云云	潍坊科技学院
6462	魏星、丁悦	内蒙古工业大学	王卓男、苏晓明	内蒙古工业大学
6477	何金田、张宝坤、姜皓宸、刘状状、李紫萱	山东工艺美术学院	温莹蕾、张静	山东工艺美术学院
6478	潘雯瑞、王佩玉、蔡倩妮、黄昊、胡惠东	潍坊学院	朱磊、韩升升	潍坊学院
6482	王佳才、何建、刘锦程	南昌大学	聂志勇	南昌大学
6489	蒋晓梅、崔奉杰、李丹阳、周万成	山东工艺美术学院	温莹蕾、张静	山东工艺美术学院
6491	戴宇奇、池笙涵、朱桐、陈相凝、彭义洁	浙江理工大学	崔艳、吴鸽鹏	浙江理工大学
6494	庄英城、王诚浩、林培标	华南理工大学	王静、肖毅强	华南理工大学
6495	张思文、陈艺丹、黄晶晶、李彤彤	厦门大学	石峰	厦门大学

续表

注册号	作者	单位名称	指导人	单位名称
6496	李雨馨、赵一凡、崔星雨	北京工业大学		
6498	郑庚锋、傅凯、金应应、邹倩	嘉兴学院	黄平、彭欣、金荣科、华昕若	嘉兴学院
6500	王靖善、韦金	北京工业大学	贾文燕、张宏然	北京工业大学
6503	凌瑞州、郑彦、姜亮	南京工业大学		
6510	林瑾如、李琴、徐寅莹、洪啸林、李孟圆	西南交通大学	李百毅	西南交通大学
6514	吴伯涛、朱赢男、陈静、周之恒、罗姜云	浙江理工大学	白文峰	浙江理工大学
6518	张雅慧、金秋彤、谢亚宁、孙瑞靖、谭辰雯、曾江倩、刘颖	北方工业大学	马欣、赵春喜	北方工业大学
6522	苏红、陈剑、冯温然、沈伟斌、吴锦凯、王嘉琪、颜君家	浙江理工大学	文强	浙江理工大学
6523	韩琪、翟利平、徐淑平、杨凯	中原工学院	孔莹博	中原工学院
6532	赵越、杨雅雯、邹元昊、李杰、王润民	华侨大学	郑志、冉茂宇	华侨大学
6533	范佳溪、朱世婷、张泽瑛、谢周诗、任丹妮、张天杰、丁奇	嘉兴学院	王荟荟、金荣科	嘉兴学院
6557	王睿涵、赵伦、李明欣、谢东、程丰睿、韩啸霖、刘晓宇、王俊峰	武汉理工大学、湖北工业大学	张辉	湖北工业大学

续表

注册号	作者	单位名称	指导人	单位名称
6561	戴芙蓉、徐涵、陈晓能	南京工业大学	杨亦陵	南京工业大学
6570	朱炜奇、王志鹏、王玥	山东建筑大学	陈兴涛	山东建筑大学建筑学院
6574	王舒恬、涂晓敏、盛晓春、赵凯	南京工业大学	陈军、林杰文	南京工业大学
6576	于洋、田园、戴卓、唐亚梅、邵婧、雷雁婷	长安大学	樊禹江、任娟	长安大学
6580	赵茂均、董鑫	重庆大学	周铁军、张海滨	重庆大学
6583	张浩、朱强华、邓汉圆、和尧熙	华中科技大学	徐燊	华中科技大学
6587	赵彬彬、田苗、费苗苗、栗月、郭文婷、张海波	西安交通大学	陈洋	西安交通大学
6589	赵与谦、陆思豪	南京工业大学	杨亦陵、张伟郁	南京工业大学
6612	陈梦雪、龚玲莉、黄婉祎	南京工业大学	邵继中	南京工业大学
6629	郑博旻、傅宇丕	长春工程学院	董峻岩	长春工程学院
6639	徐致远、杜洋	武汉大学	李鹍	武汉大学
6643	张钰、曹峡锡、丁瀚林	西南交通大学	王俊	西南交通大学
6653	何青、田毓丰、赵莹莹、赵营	苏州科技大学	邱德华、王一鸣	苏州科技大学
6654	张昕烁、伍长生	重庆大学	周铁军、张海滨	重庆大学
6661	吕一玲、祁月雨、苏泽芸	山东建筑大学	陈兴涛、李晓东	山东建筑大学

2019台达杯国际太阳能建筑设计竞赛办法
Competition Brief for International Solar Building Design Competition 2019

竞赛宗旨：

在乡村振兴战略下，如何进一步激活各种乡村资源，通过研学基地、驿站等多种载体，提升乡村发展的附加值，走好乡村绿色、可持续的发展之路。

竞赛主题：阳光·文化之旅
竞赛题目：河北兴隆暗夜公园星空驿站项目
　　　　　浙江凤溪玫瑰教育研学基地项目
主办单位：国际太阳能学会
　　　　　中国可再生能源学会
　　　　　全国高等学校建筑学学科专业指导委员会
承办单位：国家住宅与居住环境工程技术研究中心
　　　　　中国可再生能源学会太阳能建筑专业委员会
支持单位：中国建筑设计研究院有限公司
冠名单位：台达集团
媒体支持：《建筑技艺》（AT）杂志
评委会专家：崔愷：中国工程院院士、全国工程勘察设计大师、中国建筑设计研究院有限公司总建筑师。
　　　　　Deo Prasad：澳大利亚科技与工程院院士、澳大利亚勋章获得者、新南威尔士大学教授。
　　　　　杨经文：马来西亚汉沙杨建筑师事务所创始人、2016梁思成建筑奖获得者。
　　　　　林宪德：台湾绿色建筑委员会主席，台湾成功大学建筑系教授。

GOAL OF COMPETITION:

Under the strategy of revitalizing rural development, it is necessary to think about how to further activate various resources in both villages and towns, enhance added value of their development through such carriers as research and study bases as well as stations, and follow a path towards a green life and sustainable development.

THEME OF COMPETITION:

SUNSHINE AND A TOUR OF CULTURE

SUBJECT OF COMPETITION:

Subject I: Starry Sky Station in Xinglong Night Park, Hebei Province
Subject II: Fengxi Rose Education Research Study Base Project, Zhejiang Province

HOST:

International Solar Energy Society
Chinese Renewable Energy Society
National Supervision Board of Architectural Education (China)

ORGANIZER:

China National Engineering Research Center for Human Settlements
China Renewable Energy Society Solar Building Specialized Committee

COORGANIZER:

China Architectural Design & Resarch Group

仲继寿：中国可再生能源学会太阳能建筑专业委员会主任委员，中国建筑学会秘书长。

王建国：中国工程院院士、全国高等学校建筑学学科专业指导委员会主任。

庄惟敏：全国工程勘察设计大师、清华大学建筑学院院长。

黄秋平：华东建筑设计研究总院副总建筑师。

冯雅：中国建筑西南设计研究院有限公司副总工程师，中国建筑学会建筑热工与节能专业委员会副主任。

组委会成员：由主办单位、承办单位及冠名单位相关人员组成。办事机构设在中国可再生能源学会太阳能建筑专业委员会。

评比办法：

1. 由组委会审查参赛资格，并确定入围作品。
2. 由评委会评选出竞赛获奖作品。

评比标准：

1. 参赛作品须符合本竞赛"作品要求"的内容。
2. 作品应具有原创性和前瞻性，鼓励创新。
3. 作品应满足使用功能、绿色低碳、安全健康的要求，建筑技术与太阳能利用技术具有适配性。
4. 作品应充分体现太阳能利用技术对降低建筑使用能耗的作用，在经济、技术层面具有可实施性。
5. 作品评定采用百分制，分项分值见下表：

评比指标	指标说明	分值
规划与建筑设计	规划布局、环境利用与融入、功能流线、无障碍设计、建筑艺术，鼓励创新	40
被动太阳能利用技术	通过专门建筑设计与建筑构造利用太阳能的技术，鼓励创新	30
主动太阳能利用技术	通过专门设备收集、转换、传输、利用太阳能的技术，鼓励创新	10
采用的其他技术	其他绿色、低碳、安全、健康技术，鼓励创新	10
技术的可操作性	作品的可实施性，技术的经济性和普适性要求	10

TITLE SPONSOR:

Delta Group

MEDIA SUPPORT:

Architecture Technique

EXPERTS OF JUDGING PANEL:

Mr. Cui Kai, Academician of China Academy of Engineering, National Design Master and Chief Architect of China Architecture Design & Research Group.

Mr. Deo Prasad, Professor of University of New South Wales, Sydney, Australia, Asia-Pacific President of International Solar Energy Society (ISES) and Professor of Faculty of the Built Environment, the Order of Australia.

Mr. Yang Jingwen, Kenneth King Mun YEANG, President of T. R. Hamzah & Yeang Sdn. Bhd. (Malaysia), 2016 Liang Sicheng Architecture Award Winner.

Mr. Lin Xiande, Chairman of Taiwan Green Building Committee; Professor of Cheng kung University, Taiwan.

Mr. Zhong Jishou, Chief Commissioner of Special Committee of Solar Building, CRES; Secretary general of the Architectural Society of China.

Mr. Wang Jianguo, Academician of China Academy of Engineering, Director of National Supervision Board of Architectural Education (China).

Mr. Zhuang Weimin, National Design Master; Dean of School of Architecture, Tsinghua University.

Mr. Huang Qiuping, Chief Architect of East China Architecture Design Research Institute.

Mr. Feng Ya, Deputy chief Engineer of China Southwest Architectural Design and Research Institute Corp., Ltd; Deputy Director of Special Committee of Building Thermal and Energy Efficiency, Architectural Society of China.

MEMBERS OF THE ORGANIZING COMMITTEE:

It is composed by competition organizer, operator and sponsor. The administration office is a standing body in Special Committee of Solar Buildings, CRES.

APPRAISAL METHODS:

1. Organizing Committee will check up eligible entries and confirm shortlist entries.
2. Judging Panel will appraise and select out the awarded works.

APPRAISAL STANDARD:

1. The entries must meet the demands of the Competition Requirement.
2. The entries should embody originality and prospective in order to encourage innovation.

设计任务书及专业术语等附件：

附件1：河北兴隆暗夜公园星空学苑项目

附件2：浙江凤溪玫瑰教育研学基地项目

附件3：专业术语

奖项设置及奖励形式：

1. 综合奖：

一等奖作品：2名　　颁发奖杯、证书及人民币50,000元奖金（税前）；

二等奖作品：4名　　颁发奖杯、证书及人民币20,000元奖金（税前）；

三等奖作品：6名　　颁发奖杯、证书及人民币5,000元奖金（税前）；

优秀奖作品：30名　　颁发证书。

2. 优秀设计方法奖：10名　颁发证书及人民币2,000元奖金（税前）。

作品设计方法报告内容丰富充实，设计过程记录完整，设计方法新颖。

3. 技术专项奖：名额不限，颁发证书。

作品采用的技术或设计方面具有创新，实用性强。

4. 建筑创意奖：名额不限，颁发证书。

作品规划及建筑设计方面具有独特创意和先导性。

作品要求：

1. 建筑设计方面应达到方案设计深度，技术应用方面应有相关的技术图纸和指标。

2. 作品图面、文字表达清楚，数据准确。

3. 作品基本内容包括：

3.1 简要建筑方案设计说明（限200字以内），包括方案构思、太阳能综合应用技术与设计创新、技术经济指标表等。

3.2 项目的竞赛作品需进行竞赛用地范围内的规划设计，总平面图比例为1：500～1：1000（含活动场地及环境设计）。

3.3 单体设计：

能充分表达建筑与室内外环境关系的各层平面图、外立面图、剖面图，比例1：200。

能表现出技术与建筑结合的重点部位、局部详图及节点大样，比例自定；其

3. The submission works should meet the demands of usable function, green and low-carbon, and health and coziness. The building technology and solar energy technology should have adaptability to each other.

4. The submission works should play the role of reducing building energy consumption by utilization of solar energy technology and have feasibility in the aspect of economy and technology.

5. A percentile score system is adopted for the appraisal as follows:

APPRAISAL INDICATOR	EXPLANATION	SCORES
Planning and Architecture design	Urban planning design, use of environmental resource and integrating into the surroundings, functional division and streamline organization, barrier free design, architectural art. Innovation is encouraged	40
Utilization of passive solar energy technology	Use of solar energy by specific architecture and construction design. Innovation is encouraged	30
Utilization of active solar energy technology	Use of solar energy though collecting, transforming, and transmitting energy by specific equipment. Innovation is encouraged	10
Other technologies	Other technologies such as: green, low carbon, safe and healthy technologies. Innovation is encouraged	10
Operability of the technology	Feasibility, economy, and popularity of relevant technology demands	10

THE TASK BOOK OF DESIGN AND PROFESSIONAL GLOSSARY

Annex 1: Starry Sky Station in Xinglong Night Park, Hebei Province

Annex 2: Fengxi Rose Education Research Study Base Project, Zhejiang Province

Annex 3: Professional Glossary

Award Setting and Awards Form:

1. GENERAL PRIZES:

First Prize:　　　　　　　　2 winners

The Trophy Cup, Certificate and Bonus RMB 50,000 (before tax) will be awarded.

Second Prize:　　　　　　　4 winners

他相关的技术图、分析图、表等。

3.4 建筑效果表现图1～4个。

3.5 参赛者须将作品文件编排在840mm×590mm的展板区域内（统一采用竖向构图），作品张数应为4或6张。中英文统一使用黑体字。字体大小应符合下列要求：标题字高：25mm；一级标题字高：20mm；二级标题字高：15mm；图名字高：10mm；中文设计说明字高：8mm；英文设计说明字高：6mm；尺寸及标注字高：6mm。文件分辨率100dpi，格式为JPG或PDF文件。

4. 参赛者通过竞赛网页上传功能将作品递交竞赛组委会，入围作品由组委会统一编辑板眉、出图、制作展板。

5. 作品文字要求：除3.1"建筑方案设计说明"采用中英文外，其他为英文；建议使用附件3中提供的专业术语。

参赛要求：

1. 欢迎建筑设计院、高等院校、研究机构、绿色建筑部品研发生产企业等单位，组织专业人员组成竞赛小组参加竞赛。

2. 请参赛者访问www.isbdc.cn，按照规定步骤填写注册表，提交后会得到唯一的注册号，即为作品编号，一个作品对应一个注册号。提交作品时把注册号标注在每副作品的左上角，字高6mm。注册时间2018年8月22日～2018年12月31日。

3. 参赛者同意组委会公开刊登、出版、展览、应用其作品。

4. 被编入获奖作品集的作者，应配合组委会，按照出版要求对作品进行相应调整。

5. 参赛者需提交作品设计方法报告。

注意事项：

1. 参赛作品电子文档和作品设计方法报告须在2019年3月1日前提交组委会，请参赛者访问www.isbdc.cn，并上传文件，不接受其他递交方式。

2. 作品中不能出现任何与作者信息有关的标记内容，否则将视其为无效作品。

3. 组委会将及时在网上公布入选结果及评比情况，将获奖作品整理出版，并对获奖者予以表彰和奖励。

4. 获奖作品集首次出版后30日内，组委会向获奖作品的创作团队赠样书2册。

5. 竞赛活动消息发布、竞赛问题解答均可登陆竞赛网站查询。

The Trophy Cup, Certificate and Bonus RMB 20,000 (before tax) will be awarded.

Third Prize: 6 winners

The Trophy Cup, Certificate and Bonus RMB 5,000 (before tax) will be awarded.

Honorable Mention Prize: 30 winners

The Certificate will be awarded.

2. PRIZE FOR EXCELLENT DESIGN METHOD: 10 winners

The Certificate and Bonus RMB 2,000 (before tax) will be awarded.

The report of works' design method is rich and full, the design process is complete, and the design method is novel.

3. PRIZE FOR TECHNICAL EXCELLENCE WORKS:

The quota is open-ended. The Certificate will be awarded.

Prize works must be innovative with practicability in aspect of technology adopted or design.

4. PRIZE FOR ARCHITECTURAL ORIGINALITY:

The quota is open-ended. The Certificate will be awarded.

Prize works must be originally creative and advanced.

REQUIREMENTS OF THE WORK:

1. The submitted drawing sheets should meet the requirements of scheme design level and should be accompanied with relevant technical drawings and technology data.

2. Drawings and text should be expressed in clear and readable way. Mentioned data should be accurate.

3. The submitted work should include:

3.1 A project description (not exceeding 200 words) including the following factors: Schematic concept design description; Integration of solar energy technology; Innovative design; Technical and economic indicators.

3.2 Participants should provide an urban design within the outline of the site of the competition. Participants will provide a site plan (including urban context / urban design) with the scale of 1：500 or 1：1000.

3.3 Monomer Design:

Participants will provide floor plans, elevations and sections with the scale of 1：200, which can fully express the relationship between architecture and indoor and outdoor environment.

Participants should provide detailed drawings (without limitation of scale) that illustrate the integration of technology in the architectural project, as well as any other relevant elements, such as technical charts, analysis diagram, and tables.

3.4 Rendering perspective drawing (1~4).

3.5 Participants should arrange the submission into four or six exhibition panels, each 840mm×590mm in size (arranged vertically). Chinese and English font type should be both in boldface. Font height is required as follows: title with word height 25mm; first subtitle with word height 20mm; second subtitle: word height 15mm; figure title: word height 10 mm; design description word height 6mm; dimensions and labels: 6mm. File resolution: 100dpi in JPEG or PDF format.

所有权及版权声明：

参赛者提交作品之前，请详细阅读以下条款，充分理解并表示同意。

依据中国有关法律法规，凡主动提交作品的"参赛者"或"作者"，主办方认为其已经对所提交的作品版权归属作如下不可撤销声明：

1. 原创声明

参赛作品是参赛者原创作品，未侵犯任何他人的任何专利、著作权、商标权及其他知识产权；该作品未在报纸、杂志、网站及其他媒体公开发表，未申请专利或进行版权登记，未参加过其他比赛，未以任何形式进入商业渠道。参赛者保证参赛作品终身不以同一作品形式参加其他的设计比赛或转让给他方。否则，主办单位将取消其参赛、入围与获奖资格，收回奖金、奖品及并保留追究法律责任的权利。

2. 参赛作品知识产权归属

为了更广泛推广竞赛成果，所有参赛作品除作者署名权以外的全部著作权归竞赛承办单位及冠名单位所有，包括但不限于以下方式行使著作权：享有对所属竞赛作品方案进行再设计、生产、销售、展示、出版和宣传的权利；享有自行使用、授权他人使用参赛作品用于实地建设的权利。竞赛主办方对所有参赛作品拥有展示和宣传等权利。其他任何单位和个人（包括参赛者本人）未经授权不得以任何形式对作品转让、复制、转载、传播、摘编、出版、发行、许可使用等。参赛者同意竞赛承办单位及冠名单位在使用参赛作品时将对其作者予以署名，同时对作品将按出版或建设的要求作技术性处理。参赛作品均不退还。

3. 参赛者应对所提交作品的著作权承担责任，凡由于参赛作品而引发的著作权属纠纷均应由作者本人负责。

声明：

1. 参与本次竞赛的活动各方（包括参赛者、评委和组委），即表明已接受上述要求。

2. 本次竞赛的参赛者，须接受评委会的评审决定作为最终竞赛结果。

3. 组委会对竞赛活动具有最终的解释权。

4. 为维护参赛者的合法权益，主办方特提请参赛者对本办法的全部条款，特别是"所有权及版权"声明部分予以充分注意。

4. Participants should send (upload) a digital version of submission via FTP to the organizing committee, who will compile, print and make exhibition panels for shortlist works.

5. Text requirement: The submission should be in English, in addition to 3.1 "architectural design description" in English and Chinese. Participants should use the words from the Professional Glossary in Appendix 3.

PARTICIPATION REQUIREMENTS:

1. Institutes of architectural design, colleges and universities, research institutions and green building product development and manufacturing enterprises organize professionals are welcomed to form a competition group to take part in the competition.

2. Please visit the website: www.isbdc.cn. You may fill the registry according to the instruction and gain an ID of your work after submitting the registry, that is work number. One work only has one ID. The number should be indicated in the top left corner of each submission work with word height in 6mm. Registration time: 22st August, 2018 – 31st December, 2018.

3. Participants must agree that the Organizing Committee may publish, print, exhibit and apply their works in public.

4. The authors whose works are edited into the publication should cooperate with the Organizing Committee to adjust their works according to the requirements of press.

5. Participants are required to submit a Report of works'design method.

IMPORTANT CONSIDERATION:

1. Participant's digital file and works'design method report must be uploaded to the organizing committee's FTP site (www.isbdc.cn) before 1st March, 2019. Other ways will not be accepted.

2. Any mark, sign or name related to participant's identity should not appear in, on or included with submission files, otherwise the submission will be deemed invalid.

3. The Organizing Committee will publicize the process and result of the appraisal online in a timely manner, compile and publish the awarded works. The winners will be honored and awarded.

4. In 30 days after the collection of works being published, 2 books of award works will be freely presented by the Organizing Committee to the competition teams who are awarded.

5. The information concerning the competition as well as explanation about all activities may be checked and inquired in the website of the competition.

ANNOUNCEMENT ABOUT OWNERSHIP AND COPYRIGHT:

Before submitting the works, participants should carefully read following clauses, fully understand and agree with them.

附件1：
河北兴隆暗夜公园星空驿站项目

一、河北兴隆暗夜公园星空驿站项目气候条件

项目位于河北省承德市兴隆县六道河镇安营寨村暗夜公园内，北纬40°41′，东经117°25′。暗夜公园处于山谷两侧坡地上，海拔约450m，场地坡度15%~35%。

兴隆县属于暖温和中温带半湿润季风型大陆性气候，四季明显，冬长夏短。春季天气多变，骤寒骤暖，气候干旱；夏季受南方暖湿气团控制，气候温和、雨水充沛、雨量集中；秋季天高气爽；冬季受北方冷气团控制，天气寒冷干燥，雨雪稀少。年平均气温在6.5~10.3℃之间，年平均降水量688.9mm。兴隆县县境多山，气温垂直变化明显。冬季多为西北季风，夏季为东南季风。

基本气象资料

气象参数：北纬（40°41′）东经（117°25′）、测量点海拔高度450m（数据来源：NASA）。

月份	空气温度 °C	相对湿度 %	水平面日太阳辐射 kW·h/(m²·d)	大气压力 kPa	风速 m/s	土地温度 °C	月采暖度日数 °C·d	供冷度日数 °C·d
一月	-9.9	44.4	2.81	96.8	3.6	-9.6	855	0
二月	-5.5	43.7	3.69	96.7	3.7	-4.5	662	0
三月	1.9	40.1	4.80	96.2	4.0	3.8	496	1
四月	11.6	34.6	5.86	95.7	4.5	14.2	203	73
五月	18.5	38.0	6.36	95.4	4.0	21.5	43	250
六月	22.2	54.1	6.05	95.0	3.4	24.5	3	355
七月	23.3	69.8	5.22	95.0	2.8	24.6	0	410
八月	22.1	70.1	4.93	95.3	2.6	23.0	0	375
九月	17.8	56.2	4.57	95.9	2.9	18.8	37	233
十月	10.6	46.1	3.73	96.4	3.3	11.7	225	66
十一月	0.6	46.6	2.80	96.7	3.8	0.9	512	0
十二月	-6.9	45.8	2.41	96.9	3.6	-6.8	757	0
年平均数	8.9	49.1	4.44	96.0	3.5	10.2	3793	1763

According to relevant national laws and codes it is made sure by the competition sponsors that all "participants" or "authors" who have submitted their works on their own initiative have received following irrevocable announcement concerning the ownership of their works submitted:

1. Announcement of originality

The entry work of the participant is original, which does not infringe any patent, copyright, trademark and other intellectual property; it has not been published in any newspapers, periodicals, magazines, webs or other media, has not been applied for any patent or copyright, not been involved in any other competition, and not been put in any commercial channels. The participant should assure that the work has not been put in any other competition by the same work form in its whole life or legally transferred to others, otherwise, the competition sponsors will cancel the qualification of participation, being shortlisted and awarded of the participant, call back the prize and award and reserve the right of legal liability.

2. The ownership of intellectual property of the works

In order to promote competition results, the participants should relinquish copyright of all works to competition administrators and titled unit except authorship. It includes but is not limited to the exercise of copyright as follows: benefit from the right of the works on redesigning, production, selling, exhibition, publishing and publicity; benefit from the right of the works on construction for self use or accrediting to others for use. Without accreditation any organizations and individual (including authors themselves) cannot transfer, copy, reprint, promulgate, extract and edit, publish and admit to use the works by any way. Participants have to agree that competition administrators and titled unit will sign the name of authors when their works are used and the works will be treated for technical processing according to the requirements of publication and construction. All works are not returned to the author.

3. All authors must take responsibility for their copyrights of the works including all disputes of copyright caused by the works.

ANNOUNCEMENT:

1. It implies that everybody who has attended the competition activities including participants, jury members and members of the Organizing Committee has accepted all requirements mentioned above.

2. All participants must accept the appraisal of the jury as the final result of the competition.

3. The Organizing Committee reserves final right to interpret for the competition activities.

4. In order to safeguard the legitimate rights and interests of the participants, the organizers ask participants to fully pay attention to all clauses in this document, especially some clauses with blue colors.

二、兴隆暗夜公园星空驿站设计任务书

1. 项目背景

项目用地位于河北省承德市兴隆县六道河镇安营寨村暗夜公园（规划中）内，北纬40°41′，东经117°25′，距北京市中心115km、距兴隆县城30km。暗夜公园处于山谷两侧的坡地上，海拔约450m，场地坡度15%~35%，附近有商业街、低层住宅、公寓，以及滑雪场、体育公园等文化娱乐设施。

暗夜公园规划面积10.89ha，是由国家天文台参与打造的、以星空观测为主线、集专业观测和科普教育于一体的专业性主题乐园，是中国第一个国际认证的暗夜公园。公园设有世界顶级大口径专业望远镜、天文台，以及周长五百余米跨越两山山巅的圆形星桥。

2. 自然条件

项目用地内植被茂密，地表景观优美，暗夜公园沿山谷两侧展开，主要设施避开夏季洪水行经区域。

3. 基础设施

基地内基础设施完备，已建有市政自来水、排水、雨水、天然气、供电及通信系统。关于交通组织，入口以外有集中停车场截流外部车辆和解决静态交通，场地内由一条外环路串联起景区入口、互动设施、星空驿站和天文台等主要设施，园区设置内部摆渡车。

4. 竞赛场地

本项目为暗夜公园的配套设施，用地位于公园中部，环形星桥的西侧，场地坡度约为25%，详见地形图。

5. 设计要求

（1）在给定的竞赛用地范围内设计星空驿站，主要为暗夜公园的游客提供住宿服务，并为天文台工作的科学家、学者提供便利，兼有餐饮和小型会议功能。每一个酒店客房单元的建筑面积为40~50m²（含交通面积），星空驿站总建筑面积为2600m²（±5%）；

（2）项目为高端定位，布置方式可采用组团式布局或单元式布局，不宜采用集中式布局，且建筑层数不宜超过2层；

（3）项目每个酒店客房单元应设置可观测天空的室外观星场所，并具有私密性；

（4）宜尽量减少建筑结构对原始地形的影响，并与周围环境有机结合；

（5）场地内的交通设计应结合地形地貌，与自然景观相结合；

（6）充分结合当地的气候特点和园内建筑特色，以及酒店用能特征，合理选择和应用主、被动太阳能技术及其他可再生能源技术，解决酒店冬季集热采暖和夏季通风降温问题，并考虑技术的可实施性。

（7）酒店功能要求如下表所示：

Annex1:
Starry Sky Station in Xinglong Night Park, Hebei Province

Climate of Starry Sky Station Project in Xinglong Night Park, Hebei Province

The project site is located in a night park of Anying Zhai Village, Xinglong County, Chengde City, Hebei Province. Its latitude and longitude are 40°41′ N and 117°25′ E respectively. The night park is seated on a sloping land along both sides of valleys with an altitude of about 450 meters and the slope ranging from 15% to 35%.

Xinglong County boasts semi-humid continental monsoon climate of warm temperate and mid-temperate zones with four distinct seasons. It lasts a long period in winter while a short one in summer. The weather in spring is changeable featuring sudden cold and warm, and it is rather dry. Influenced by the southern warm and moist air mass, Xinglong County in summer is characterized by mild climate, abundant rain and concentrated precipitation. It is much cool and refreshing in autumn. Besides, influenced by the cold air mass in the north as well, it is cold and dry in winter with rare rain and snow. The annual average temperature fluctuates between 6.5 °C and 10.3 °C, and the average annual precipitation reaches 688.9 millimeters. There lie endless mountains within Xinglong County, which presents a distinct vertical change in temperature. Northwest monsoon mostly falls in winter, while southeast monsoon falls in summer.

Basic meteorological data are shown in the attached table.

Meteorological parameters: 40°41′N, 117°25′E, 450m of altitude of the measuring point (data from NASA)

Month	Temperature °C	Relative Humidity %	Level Solar Radiation kW·h/(m²·d)	Air Pressure kPa	Wind Speed m/s	Ground Temperature °C	Gross Heating Degree °C·d	Gross Cooling Degree °C·d
JAN	-9.9	44.4	2.81	96.8	3.6	-9.6	855	0
FEB	-5.5	43.7	3.69	96.7	3.7	-4.5	662	0
MAR	1.9	40.1	4.80	96.2	4.0	3.8	496	1
APR	11.6	34.6	5.86	95.7	4.5	14.2	203	73
MAY	18.5	38.0	6.36	95.4	4.0	21.5	43	250
JUN	22.2	54.1	6.05	95.0	3.4	24.5	3	355
JUL	23.3	69.8	5.22	95.0	2.8	24.6	0	410
AUG	22.1	70.1	4.93	95.3	2.6	23.0	0	375
SEP	17.8	56.2	4.57	95.9	2.9	18.8	37	233
OCT	10.6	46.1	3.73	96.4	3.3	11.7	225	66
NOV	0.6	46.6	2.80	96.7	3.8	0.9	512	0
DEC	-6.9	45.8	2.41	96.9	3.6	-6.8	757	0
YEAR	8.9	49.1	4.44	96.0	3.5	10.2	3793	1763

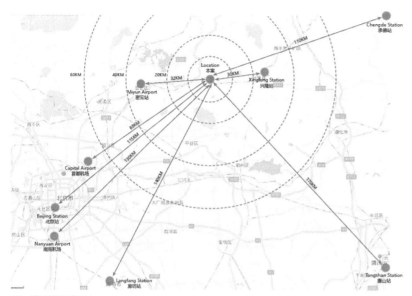

图1 项目所在区位图
Figure 1　Bitmap of the Project Site

图2　暗夜公园项目所在地周边环境
Figure 2　Surrounding Environment of the Night Park Project Site

图3　暗夜公园效果图
Figure 3　Effect Picture of the Night Park

Design Task Book of Starry Sky Station in Xinglong Night Park, Hebei Province

1. Project Background

The project site is located in a night park (under planning) of Anying Zhai Village, Liudaohe Town, Xinglong County, Chengde City, Hebei Province. Its latitude and longitude are 40°41′N and 117°25′E respectively. The site is 115 kilometers away from the Beijing city center and 30 kilometers away from Xinglong County. The night park is located on the slope alongside the valley with an altitude of about 450 meters and a slope ranging from 15% to 35%. There are commercial streets, low-rise residential buildings and apartments, and cultural and recreational facilities such as ski resorts and sports parks nearby.

The night park with a planned area of 10.89 hectares is a professional theme park jointly built by the National Astronomical Observatories with starry sky observation as its principal line and integrating professional observation and popular science education. It is the first internationally authenticated night park in China, equipped with the world's top and professional telescopes with heavy caliber, observatories and a circular star bridge with a circumference of over 500 meters spanning two mountain peaks.

2. Natural Condition

The project site is densely covered with vegetation and boasts a beautiful surface landscape. The night park spreads along both sides of the valley whose main facilities avoid the areas where floods pass by in summer.

3. Infrastructure

The base has been equipped with self-contained infrastructures, and is accessible to systems of municipal tap water, drainage, natural gas rainwater, power supply and communications. In respect to traffic organization, there are centralized parking lots outside the entrance to intercept external vehicles and solve the problem of static traffic, and the site is connected by an outer ring road with major facilities such as scenic spot entrance, interactive facilities, starry sky stations and observatories. Besides, ferry buses are equipped in the park internally.

4. Competition Venues

This project constitutes a supporting facility for the night park. The project land is located in the middle of the park and on the west of the ring-shaped star bridge. The slope grade of the site is about 25 %. Details can be obtained from the topographic map.

5. Design Requirements

(1) Starry Sky Stations are designed within the given competition area, mainly providing accommodation services for night park tourists, and providing convenience for scientists and scholars working at the observatory, as well as exhibiting the functions for catering and small conferences. The construction area of each hotel room unit reaches 40 to 50m² (traffic area included), and the total construction area of the starry sky station is about 2,600m² (more than or less than 5%);

(2) The project is positioned for high-end use. The layout can be arranged by a group-based method or unit-based method, instead of a method of adopting centralized layout. Moreover, the number of floors of the building should not

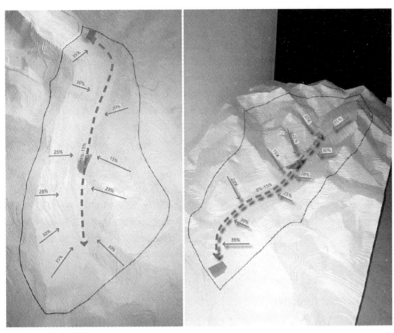

图4 项目用地坡度和坡向观
Figure 4 Slope Grade and Aspect of the Project Land

exceed two.

(3) Outdoor star observation venues should be arranged in each guest room of the project, and privacy of the rooms should be guaranteed.

(4) The influence of the building structure on original topography should be minimized as far as possible, and the project should be integrated with the surrounding environment organically.

(5) Traffic design in the venue should be integrated with natural landscape based on geographic and geomorphic conditions.

(6) Active and passive solar energy and other renewable energy application technologies are rationally selected and applied in terms of the combination of local climate characteristics, architectural features inside the park, features of energy consumption of hotels, so as to solve the problems of heat collection in winter and ventilation for cooling in summer of hotels. In addition, the economy of technologies should be considered as well.

(7) The functional requirements of the hotel are displayed in the following table:

	酒店功能要求	数量（间）	面积（m²）	备注
客房部分	客房单元	≥25	1000~1250	
	客房服务用房	2		
公共部分	酒店大堂	1	400~450	
	大堂服务配套	1		
	会议或活动厅	1~2		可容纳50人
餐饮部分	厨房	1~2	300~350	
	餐厅	1~2		
	酒吧（大堂）	1~2		
行政部分	办公室		400~450	
后勤部分	后勤服务用房			

	Functional Requirements of the Hotel	Quantity (room)	Area (m²)	Remarks
Room	Guest room	≥25	1000~1250	
	Room for guest services	2		
Common area	Hotel lobby	1	400~450	
	Supporting room for lobby services	1		
	Conference or event hall	1~2		accommodating 50 people.
Catering	Kitchen	1~2	300~350	
	Restaurant	1~2		
	Bar (lobby)	1~2		
Administration	Office		400~450	
Logistics	Room for logistics services			

附件 2：
浙江凤溪玫瑰教育研学基地项目

一、浙江凤溪玫瑰教育研学基地项目气候条件

项目位于浙江省杭州市桐庐县凤川街道南部三鑫村凤溪玫瑰教育研学基地内，北纬29°47′，东经119°10′。项目处于三山夹谷、两溪交汇的小河谷平原，海拔约85m。

桐庐县属于亚热带季风气候，四季分明，日照充足，降水充沛。一年四季光、温、水基本同步增减，配合良好，气候资源丰富。年平均气温约16℃，年平均降水量约为1552mm，6月为梅雨期，降水较为集中。

基本气象资料

气象参数：北纬（29°47′）东经（119°10′）、测量点海拔高度85m（数据来源：NASA）。

月份	空气温度 ℃	相对湿度 %	水平面日太阳辐射 kW·h/(m²·d)	大气压力 kPa	风速 m/s	土地温度 ℃	月采暖度日数 ℃·d	供冷度日数 ℃·d
一月	4.4	73.4	2.49	99.4	3.4	4.5	415	3
二月	6.0	74.0	2.59	99.2	3.4	6.5	333	11
三月	9.7	75.7	2.91	99.8	3.2	10.3	257	42
四月	15.4	76.5	3.73	98.4	3.1	16.2	98	162
五月	20.0	76.4	4.24	98.0	2.8	20.8	16	306
六月	23.6	79.7	4.07	97.6	2.9	24.3	0	406
七月	26.5	79.8	5.15	97.5	2.8	27.0	0	517
八月	25.7	81.6	4.63	97.6	2.8	26.3	0	492
九月	22.0	80.2	3.84	98.2	3.2	22.7	1	366
十月	17.2	74.1	3.40	98.8	3.2	17.9	45	230
十一月	12.0	71.6	2.90	99.2	3.4	12.4	173	89
十二月	6.6	69.9	2.76	99.5	3.3	6.6	343	15
年平均数	15.8	76.1	3.56	98.5	3.1	16.3	1681	2639

二、浙江凤溪玫瑰教育研学基地项目设计任务书

1. 项目背景

项目用地位于浙江省杭州市桐庐县凤川街道南部三鑫村凤溪玫瑰教育研学基地内，北纬29°47′，东经119°10′，项目地处三山夹谷、两溪交汇的小河谷平原，距G25高速凤川出口3km，距杭州市区70km。

Annex2：
Fengxi Rose Education Research Study Base Project, Zhejiang Province

Climate of Fengxi Rose Education Research Study Base Project, Zhejiang Province

The project is located in the Fengxi Rose Education Research Study Base in Sanxin Village, Fengchuang Avenue South, Tonglu County, Hangzhou City, Zhejiang Province. It is 29°47′N and 119°47′E in latitude and longitude respectively. Seated in a small valley plain besieged with three mountains and two streams, the project reaches about 85 meters in the altitude.

Tonglu County boasts a subtropical monsoon climate with four distinct seasons, ample sunshine and abundant rainfall. There exists a sound coordination among light, temperature and rainfall, all of which maintain basic synchronization in the increase and decrease, providing abundant climate resources. The annual average temperature is 16.5℃, and the annual average precipitation is 1,552 mm. The Meiyu period falls on June, resulting in the most intensive rainfall.

Basic Meteorological Data

Meteorological parameters: 29°47′N, 119°10′E, 85m of altitude of the measuring point (data from NASA)

Month	Temperature ℃	Relative humidity %	Level Solar Radiation Per Day kW·h/(m²·d)	Air Pressure kPa	Wind Speed m/s	Ground Temperature ℃	Monthly Heating Degrees ℃·d	Cooling Degree Days ℃·d
JAN	4.4	73.4	2.49	99.4	3.4	4.5	415	3
FEB	6.0	74.0	2.59	99.2	3.4	6.5	333	11
MAR	9.7	75.7	2.91	99.8	3.2	10.3	257	42
APR	15.4	76.5	3.73	98.4	3.1	16.2	98	162
MAY	20.0	76.4	4.24	98.0	2.8	20.8	16	306
JUN	23.6	79.7	4.07	97.6	2.9	24.3	0	406
JUL	26.5	79.8	5.15	97.5	2.8	27.0	0	517
AUG	25.7	81.6	4.63	97.6	2.8	26.3	0	492
SEP	22.0	80.2	3.84	98.2	3.2	22.7	1	366
OCT	17.2	74.1	3.40	98.8	3.2	17.9	45	230
NOV	12.0	71.6	2.90	99.2	3.4	12.4	173	89
DEC	6.6	69.9	2.76	99.5	3.3	6.6	343	15
YEAR	15.8	76.1	3.56	98.5	3.1	16.3	1681	2639

Design Task Book of Fengxi Education Research Study Base in Zhejiang Province

1. Project Background

The project is located in the Fengxi Rose Education Research Study Base in Sanxin Village, Fengchuang Avenue South, Tonglu County, Hangzhou

浙江凤溪玫瑰教育研学基地项目一期已建成玫瑰农场、环艺展示空间、美食体验空间、手工制作空间、民宿体验营地等配套设施，二期将建设教学中心、创意中心、学生宿舍、体验单元等。本项目结合研学基地的建设需求，充分利用太阳能等可再生能源技术，结合周边优越的自然环境，建设适用于夏热冬冷地区的绿色、低碳、健康的研学服务设施。

2. 自然条件
项目用地内植被茂密，地表景观优美，南侧靠山，西北临小溪，东北侧临公路。

3. 基础设施
基地内基础设施完备，已建有自来水、排水、雨水、供电及通信系统。

4. 竞赛场地
本项目为凤溪玫瑰教育研学基地二期项目，用地位于园区西侧，南高北低，高差3m。场地内有两处水塘，位置见地形图。

City, Zhejiang Province with the latitude and longitude of 29°47'N and 119°47'E respectively. Seated in a small valley plain besieged with three mountains and two streams, it is three kilometers away from the Fengchuan Exit of G25 Expressway, and 70 kilometers away from downtown Hangzhou.

Such supporting facilities as the Rose Farm, Environmental Art Exhibition Space, Gourmet Experience Space, Handcraft Space and Homestay Experience Camp have been completed in the Fengxi Rose Education Research Study Base in Zhejiang. The teaching center, creative center, students' dormitory and experience units will be built in the second phase. In consideration of construction demands of the research study base and superior natural environment around it, the project makes full use of renewable energy technologies, such as solar energy to establish green, low-carbon and healthy research study service facilities suitable for regions where it's hot in summer and cold in winter.

2. Natural Conditions
The project site is densely covered with vegetation and boasts a beautiful surface landscape. The project is adjacent to mountains in the south, creek in the northwast, and highway in the northeast.

3. Infrastructure
The base has been equipped with complete infrastructures, and is accessible to systems of tap water, drainage, rainwater, power supply and communications.

4. Competition Venues
This project constitutes the second phase project of the Fengxi Rose Education Research Study Base. The land is located on the wast of the park, which is higher in the south and lower in the north with a three-meter height difference. In the site there exist two reservoirs whose locations are shown in the topographic map.

5. Design Requirements
(1) The teaching center, creative center, research dormitory and experience units are designed within the given competition area, covering a total construction area of about 2,800 square meters. (more than or less than 5%);

图1 项目所在区位图
Figure 1 Bitmap of the Project Site

图2 项目用地地形及地表景观1
Figure 2 Terrain and Surface Landscape of the Project Site 1

(2) The teaching center is built for students to study related courses, with one floor covering an area of 700 square meters.

(3) The creative center is cultural and creative space for creation and exchanges of creative teams and volunteers, is also a core working area for management teams with two floors covering a construction area of 550 square meters. There are two floors in total in the building. Among them, the independent creation space covers many different themes, such as themes of music creation, gardening (the additional sunlight greenhouse area is unlimited, excluding from the total construction area), clothing and fabrics, painting and children's paradise.

图3 项目用地地形及地表景观2
Figure 3 Terrain and Surface Landscape of the Project Site 2

图4 项目现场航拍图
Figure 4 Aerial Image of the Project Site

(4) The research study dormitory is a place that provides accommodation for research study teachers, students and parents. it has one floor or two floors covering a construction area of 750-900 square meters, which is appropriate to be divided into five or six groups.

图5 凤溪玫瑰研学基地入口
Figure 5 Entrance of the Fengxi Rose Education Research Study Base

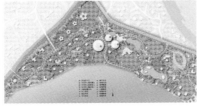
图6 凤溪玫瑰教育研学基地规划示意图
Figure 6 Planning Graph of the Fengxi Rose Education Research Study Base

(5) The experience unit provides family-based experience space for

5. 设计要求

（1）在给定的竞赛用地范围内设计教学中心、创意中心、研学宿舍、体验单元，总建筑面积为 2800m²（±5%）；

（2）教学中心为学生研学相关课程的授课空间，建筑面积 700m²，建筑层数 1 层；

（3）创意中心是一个文化创意空间，为创意团队和志愿者提供创作场所和交流空间，也是管理团队的核心工作区，建筑面积 550m²，建筑层数 2 层。其中独立创作空间为：音乐主题、园艺主题（附加阳光温室面积不限，不计入总建筑面积）、服装布艺主题、绘画主题、儿童乐园等空间；

（4）研学宿舍是为研学教师、学生及家长提供住宿的场所，建筑面积 750~900m²，建筑层数 1~2 层，宜分 5~6 组独立单元；

（5）体验单元为研学学生、家长及游客提供以家庭为单位的体验空间，每个单元建筑占地 700m²、建筑面积≤100m²，建筑层数 2 层；

（6）充分结合当地的气候特点和乡村田园的建筑特色，以及不同建筑的用能特征，解决建筑的冬季集热采暖和夏季通风降温问题，合理选择和应用主、被动太阳能技术及其他可再生能源技术，并考虑技术的可实施性。

（7）建筑用房设置如下表所示：

	功能要求	数量	面积（m²）	备注
教学中心	小型教室	6 个	30m²/个	
	中型教室	2 个	70~80m²/个	
	大型教室	1 个	200m²/个	
	功能实验室	3 个	30~50m²/个	
	卫生间		20m²	
创意中心	共享创作空间		200m²	
	独立创作空间	4 个	30m²/个	含专用卫生间
	餐饮休息空间		80m²	
	儿童乐园		120m²	
	厨房		60m²	
	卫生间		10~20m²	
研学宿舍	研学宿舍	30 间	30~40m²/个	每个宿舍 4 人，有独立卫生间
体验单元	体验单元	3 个	建筑占地 700m²/个，建筑面积≤100m²/个，基底面积 50m²/个	每个体验单元可容纳 4~6 人（3 代家庭），建筑层数两层

students, parents and tourists, with each unit covering an area of 700 square meters, and the unit has two floors covering a construction area of less than or equal to 100 square meters.

(6) Active and passive solar energy and other renewable energy application technologies are rationally selected and applied in terms of the combination of local climate characteristics, architectural features inside the park, features of energy consumption of hotels, so as to solve the problems of heat collection in winter and ventilation for cooling in summer of hotels. In addition, the economy of technologies should be considered as well.

(7) The construction room setting is shown in the following table:

	Functional Requirements	Quantity	Area（m²）	Remarks
Teaching Center	Small classrooms	6	30m² per classroom	
	Medium-sized classrooms	2	70~80m² per classroom	
	Large classrooms	1	200m² per classroom	
	Functional laboratories	3	30~50m² per laboratory	
	Toilet		20m²	
Creative Center	Shared creative space		200m²	
	Shared creative space	4	30m² per space	Particular toilets included
	Rest space of food and beverage		80m²	
	Children's paradise		120m²	
	Kitchen		60m²	
	Toilet		10~20m²	
Research Study Dormitory	Research study dormitory	30 rooms	30~40m² per room	Four people in each dormitory, with an independent toilet
Experience Unit	Experience units	3	It covers an area of 700 square meters, with the construction area of less than or equal to 100 square meters, and the base area is 50 square meters per unit	Each experience unit can accommodate 4~6 people (three generations of families), with two floors of buildings

附件 3：／Annex 3：专业术语 Professional Glossary

百叶通风	shutter ventilation
保温	thermal insulation
被动太阳能利用	passive solar energy utilization
敞开系统	open system
除湿系统	dehumidification system
储热器	thermal storage
储水量	water storage capacity
穿堂风	through-draught
窗墙面积比	area ratio of window to wall
次入口	secondary entrance
导热系数	thermal conductivity
低能耗	lower energy consumption
低温热水地板辐射供暖	low temperature hot water floor radiant heating
地板辐射采暖	floor panel heating
地面层	ground layer
额定工作压力	nominal working pressure
防潮层	wetproof layer
防冻	freeze protection
防水层	waterproof layer
分户热计量	household-based heat metering
分离式系统	remote storage system
风速分布	wind speed distribution
封闭系统	closed system
辅助热源	auxiliary thermal source
辅助入口	accessory entrance
隔热层	heat insulating layer
隔热窗户	heat insulation window
跟踪集热器	tracking collector
光伏发电系统	photovoltaic system
光伏幕墙	PV façade
回流系统	drainback system
回收年限	payback time
集热器瞬时效率	instantaneous collector efficiency
集热器阵列	collector array
集中供暖	central heating
间接系统	indirect system
建筑节能率	building energy saving rate
建筑密度	building density
建筑面积	building area
建筑物耗热量指标	index of building heat loss
节能措施	energy saving method
节能量	quantity of energy saving
紧凑式太阳热水器	close-coupled solar water heater
经济分析	economic analysis
卷帘外遮阳系统	roller shutter sun shading system
空气集热器	air collector
空气质量检测	air quality test (AQT)
立体绿化	tridimensional virescence
绿地率	greening rate
毛细管辐射	capillary radiation
木工修理室	repairing room for woodworker
耐用指标	permanent index
能量储存和回收系统	energy storage & heat recovery system
平屋面	plane roof
坡屋面	sloping roof
强制循环系统	forced circulation system
热泵供暖	heat pump heat supply
热量计量装置	heat metering device
热稳定性	thermal stability
热效率曲线	thermal efficiency curve
热压	thermal pressure
人工湿地效应	artificial marsh effect
日照标准	insolation standard
容积率	floor area ratio
三联供	triple co-generation
设计使用年限	design working life
使用面积	usable area
室内舒适度	indoor comfort level

双层幕墙	double façade building
太阳方位角	solar azimuth
太阳房	solar house
太阳辐射热	solar radiant heat
太阳辐射热吸收系数	absorptance for solar radiation
太阳高度角	solar altitude
太阳能保证率	solar fraction
太阳能带辅助热源系统	solar plus supplementary system
太阳能电池	solar cell
太阳能集热器	solar collector
太阳能驱动吸附式制冷	solar driven desiccant evaporative cooling
太阳能驱动吸收式制冷	solar driven absorption cooling
太阳能热水器	solar water heating
太阳能烟囱	solar chimney
太阳能预热系统	solar preheat system
太阳墙	solar wall
填充层	fill up layer
通风模拟	ventilation simulation
外窗隔热系统	external windows insulation system
温差控制器	differential temperature controller
屋顶植被	roof planting
屋面隔热系统	roof insulation system
相变材料	phase change material (PCM)
相变太阳能系统	phase change solar system
相变蓄热	phase change thermal storage
蓄热特性	thermal storage characteristic
雨水收集	rain water collection
运动场地	schoolyard
遮阳系数	sunshading coefficient
直接系统	direct system
值班室	duty room
智能建筑控制系统	building intelligent control system
中庭采光	atrium lighting
主入口	main entrance
贮热水箱	heat storage tank
准备室	preparation room
准稳态	quasi-steady state
自然通风	natural ventilation
自然循环系统	natural circulation system
自行车棚	bike parking